Cost Control

for Builders, Remodelers, and Developers

Jerry Householder

Home Builder Press ®
National Association of Home Builders
1201 15th St., N.W.
Washington, D.C. 20005-2800

Cost Control for Builders, Remodelers, and Developers

ISBN 0-86718-405-1

Printed in the United States of America.

Library of Congress Cataloging-in Publication Data

Householder, Jerry
Cost Control for builders, remodelers, and developers / Jerry Householder.
p. cm.
Includes bibliographical references.
ISBN 0-86718-405-1 : $25.00
1. Building—Cost control. 2. Construction industry—Management. I. Title.
TH437.H72 1994
690'.0681—dc20 94-15859
CIP

For further information, please contact—
Home Builder Press®
National Association of Home Builders
1201 15th Street, NW
Washington, DC 20005-2800

12/94 Edington-Rand/McN & Gunn 3000

Acknowledgments

The author would like to thank the many people who contributed directly and indirectly for their assistance in the formation of this book including Walter R. Bankston, a former member of the Board of Directors of the NAHB National Council of the Housing Industry; Iris Maughan, a builder and remodeler with special talents, Billy Coates, whose cost control philosophy was most influential, and Rollin Rockett, who has forgotten more about cost control than most of us will ever know.

Thanks also to the reviewers Alan Hanbury, CGR, Treasurer, House of Hanbury, Newington, Connecticut; Robert Hankin, Vice President, RDR Construction Company, Highland Mills, New York; Richard McCleery, President, John Morley, Ltd., Lexington, Ohio; Joseph R. Molinaro, Director, Land Development Services, National Association of Home Builders (NAHB); Chuck Moriarty, CGR, President, Moriarty and Matzen, Seattle, Washington; John Mouton, Chairman, Department of Building Science, Auburn University, Auburn, Alabama; Joseph W. Schwartz, President, J-II Homes, Inc., Fairfield, Ohio; and Glen Williams, President/ Senior Modeler, Envision Studio, Harrisburg, Pennsylvania.

Book Preparation

Cost-Control for Builders, Remodelers, and Developers was produced under the direction of Kent Colton, NAHB Executive Vice President, in association with NAHB staff members Jim Delizia, Staff Vice President for Member and Association Relations; Adrienne Ash, Assistant Vice President, Publishing Services; Rosanne O'Connor, Director of Publications; Doris M. Tennyson, Director, Special Projects/Senior Editor and Project Editor; Julie Wilson, Marketing Director; David Rhodes, Art Director; and Carolyn Poindexter, Editorial Assistant.

Note

Some builders, remodelers, and developers use the term *indirect costs* to refer to costs for such items as insurance and permits that should be charged to a project but will not be built into the project like 2x4s and concrete. The author of this book has chosen to use the term *job overhead* instead. In researching over a thousand court cases he found that in every case the courts frequently used the term *overhead* but none of them used *indirect costs*.

Contents

Acknowledgments ii
Introduction 1

1. Are Your Costs Under Control? 5
Benefits of Cost Control 5
Goals of Cost Control 5
Factors Affecting Record Keeping 7
External Factors Affecting Cost Control 12
Internal Factors Affecting Cost Control 14
Ways to Keep Records 16

2. Plan Your Approach to Cost Control 18
Job Costs 18
General Overhead 27

3. Track and Analyze Job Costs 31
Set Up Cost Categories 31
Establish a Budget 32
Post Ledgers 43
Computerize Cost Control 52
Make Operational Decisions 56
Develop Estimating Data base 56
Enhance Cost Effectiveness 57

4. Track and Analyze General Overhead Costs 58
Set Up Cost Categories 58
Post the General Overhead Ledger 63
Analyze General Overhead Expense 63

5. Operate Your Purchasing System 67
Materials 67
Purchasing Objectives 68
Subcontractors 72
Employees 76
Equipment 80
Customer-Related Costs 81

6. Track and Analyze Remodeling Costs 85
Budget for a Cost-Plus or a Fixed-Price Contract 85
Change Orders 85

7. Track and Analyze Development Costs 96
Establish a Budget 98
Set Up Cost Centers 98
Post Ledgers 98
Variance Between Budget Estimate and Cost 103

Notes 105
About the Author 107

Figures

Chapter 1. Are Your Costs Under Control?
1-1. Factors that Impact Productivity 8

Chapter 2. Plan Your Approach to Cost Control
2-1. Checklist for Planning Your Cost Control System 30

Chapter 3. Track and Analyze Job Costs
3-1. Sample Floor Plan and Elevation 33
3-2. Sample Estimate 34
3-3. Sample Budget for Single-Family House 43
3-4. Sample Job Cost Ledger 44
3-5. Method of Doing Variance Analyses 51
3-6. Sample Spreadsheet Cost Control System 53
3-7. NAHB Approved Software® Seal 56
3-8. Checklist to Track and Analyze Costs 57

Chapter 4. Track and Analyze General Overhead Costs
4-1. Manually Posted Sample General Overhead Ledger 64
4-2. Checklist for Tracking and Analyzing General Overhead 66

Chapter 5. Operate Your Purchasing System
5-1. Checklist for Operating Your Purchasing System 68
5-2. Sample Purchase Order 71
5-3. Reasons for Using Subcontractors 73
5-4. Questions to Ask Yourself About Renting, Leasing, or Buying Equipment 80

Chapter 6. Track and Analyze Remodeling Costs
6-1. Sample Cover Letter for Change Order for Extra Exploratory Work 86
6-2. Sample Change Order for Extra Exploratory Work 87
6-3. Sample Cover Letter for Change Order for Extra Work 88
6-4. Activities in the Scope of the Work 91
6-5. Recap of Estimate 92
6-6. Sample Job Ledger 93
6-7. Tracking and Analyzing Remodeling Costs 92

Chapter 7. Track and Analyze Development Costs
7-1. Businesses That Profit from Land Development, Construction, and Sale of a House 97
7-2. Sample Land Development Costs Subsidiary Ledger 99
7-3. Sample Manually Posted Job Cost Ledger for Developers 100
7-4. Sample Computerized Job Cost Ledger for Developers 102
7-5. Checklist for Tracking and Analyzing Development Costs 104

Introduction

If you ask 10 builders, remodelers, or developers to define *cost control,* you will probably get 10 different answers. Just about everything a builder does in building a house from choosing a material supplier to inspecting a subcontractor's work may be considered to be cost control measures. The same is true for a remodeler building an addition or a developer producing finished lots.

The building business has many specialists. Some builders build primarily speculative homes. Some specialize in custom or contract homes. Other builders diversify into remodeling. Many remodelers do only remodeling, but others also do some building. Developers may stick only to development, or they may build too. Your cost control system should mirror your particular business.

Cost Control for Builders, Remodelers, and Developers considers two broad functions of the management of a builder's, remodeler's, or developer's business—record keeping and purchasing procedures. It shows how to set up and benefit from record-keeping procedures and purchasing systems used by other successful builders, remodelers, and developers. However the book will not ignore the business management costs, which also must be controlled. To control costs, you need to understand how to allocate these costs and make sure they are absorbed by the individual projects.

No construction, remodeling, or development company is so small that it cannot profitably apply the accepted methods of business organization. One important method is to follow a business plan. The first four crucial steps you need to take in formulating a company business plan include—

- the establishment of a mission statement that is the fundamental purpose of the business
- define the general scope of the company's operations
- establish long-term objectives
- define a plan of action to achieve the goals

For example, a builder may determine that an appropriate mission statement is the production of high-quality homes built in a professional, businesslike manner. The general scope of the company's operation may be to provide affordable homes for first-time homebuyers. The long-term objectives may include the construction of 20 homes during the first year and 30 homes per year thereafter. The plan of action might include the items listed below:

- Start one home every $2\frac{1}{2}$ weeks for the first year.
- Determine where you can buy lots.
- Establish the general procedures for record keeping, purchasing, and associated business requirements, including a cost control system.
- Obtain the financing and arrange to buy 10 lots.
- Hire office and field personnel to maintain the planned production.
- Establish policies and procedures for handling changes in the scope of the work.

A remodeler may determine that a suitable mission statement is the remodeling of mid-price-range homes for owners who prefer to remain in their existing homes but need or want more or better space. The scope of the company's operation may be to specialize in the creation of new space outside the bounds of the existing home. Another remodeler's mission statement might also be to provide a reliable alternative to homeowners who want a larger or better place to live, but the general scope of the operation may be to concentrate on remodeling homes over 75 years old.

The long-term objectives for both remodelers could include (a) do work of such quality that they will always have a backlog of work and (b) gradually work up to a gross sales volume of $1 million dollars of work per year. Their action plans might include the tasks listed below:

- Develop reliable, competent craftsmen and subcontractors.
- Formulate an appropriate marketing strategy.
- Establish business procedures for record keeping, purchasing, and other management functions, including tracking and analyzing costs.
- Establish banking or financing relationships.
- Seek out sources of hard-to-find materials.
- Establish policies and procedures for handling changes in the scope of the work.
- Set up policies and procedures for dealing with hazardous materials when a job involves demolition.

A developer's mission statement might be to produce building lots in response to market needs. The scope of the company's operation might be limited to a certain geographical area and a definite subdivision size. Long-term objectives might include expansion or attaining a certain average annual volume. The plan of action might encompass the items listed below:

- Develop a reliable market survey strategy.
- Determine the best marketing methodology.
- Establish banking or financing relationships.
- Establish appropriate business procedures.

Running a successful building, remodeling, or land development business involves a number of separate, unrelated activities such as marketing, estimating, building, negotiating, choosing or designing plans, choosing locations (for contract or speculative houses), planning and scheduling, purchasing, material management, project management, accounting, cost control, customer service, and risk management (including safety and insurance). To succeed in business, a builder, remodeler, or developer should generate a business plan to help pull together these activities in a structured way. In formulating your business plan, you may want to consider all of these activities.

Cost control is defined as "a monitoring process that provides feedback to the manager concerning project expenses and how they compare to the established budget."[1] Cost control has to be included in the business plan that is based on the mission statement. It is a way of operating a business, an integral part of the operation and not something separate. A cost control system includes all of the activities builders, remodelers, and developers use to make the most profit from their work while improving their competitive position in the market.

In instituting any control system you must be able to justify the expense of the system with the value of the data it provides. Such a system has little or no use if the information provided is not in a usable form or is not produced in a timely manner. The builder, remodeler, or developer must look at each control system as an investment that yields a fair return on the cost of the system. If the investment is not yielding the proper return, you need to evaluate the structure or the use of the system or both.

The basic cost-tracking system outlined in this book has been used successfully by many builders, remodelers, and developers. It offers many rewards for a small investment of time and effort. The information gained from this basic system is used to develop cost data that will be extremely useful in (a) estimating the cost of future work and (b) keeping the cost of ongoing projects within the established budgets.

The builder, remodeler, or developer should prepare a detailed estimate for each home, remodeling job, or lot. (See *Estimating for Home Builders*[2]). This estimate serves, not only as the basis for a bid for contract work and a target estimate for speculative work, but also as a budget for cost control during the construction of the project.

As project expenses occur they are charged against the particular part of the project to which they apply. You need to continuously compare these costs with the budget to determine exactly where you are in terms of cost compared to where you planned to be.

Not knowing exactly where you stand financially on a project is inexcusable, especially when doing so is so easy. To maximize your profits is extremely difficult if you do not keep within established budgets and do not know when job costs are excessive. The only way to keep field expenses from becoming excessive is to be able to isolate the exact causes.

Simply knowing that you are bleeding to death financially is of little use if you do not know from where the blood is coming.

Cost control requires a systematic effort of planning, design, purchasing, production management, job cost accounting, post construction warranty, and review. The approach to job cost control presented in these pages is only one of many possible approaches. Some other approaches are more rudimentary and are therefore easier but do not yield the kind of data many builders, remodelers, or developers need. Still other methods of cost control are more sophisticated than those presented here. In adopting this approach, you need to tailor it to your particular business circumstances.

One

Are Your Costs Under Control?

How well does your business control costs? If your cost control system is working properly, you should be able to tell if you are making or losing money on a project while that project is in progress. To find out if you are making money, you must have a budget established for each major phase of the project. You also must have a method of tracking your costs as they are incurred and a way to compare these costs to the budget. To establish a budget for a project requires a detailed estimate of the costs for that project regardless of whether it is a contract or speculative home, a land development project, or a remodeling job. Unless you have a realistic budget for the major work items, you cannot know for sure if you are ahead or behind financially.

Benefits of Cost Control

How quickly do you react to the cost control information you generate? If you establish a budget for each major phase of the work and track the costs, if you record or post the costs in a timely manner, and if you review the data on a regular basis, you should be able to take the proper action to keep unjustified expenses to a minimum.

One builder who instituted a cost control system similar to the one proposed in this book has compared his business before nstituting cost control to trying to do a jigsaw puzzle in the dark. He reports that using the cost control system was like turning on a light.

Goals of Cost Control

Listed below are five goals of cost control followed by brief descriptions of how to achieve them:

- increase profits
- improve efficiency and quality
- improve employee morale and loyalty
- increase productivity
- reduce waste

Increase profits—You have many reasons to integrate cost control measures into your business plan. Profits can increase if you learn (in time to take corrective actions) that one phase of a project is costing more than it should. Likewise, if you have reliable historical data, you can analyze alternatives to determine the cost effectiveness of different choices. For example, in many areas potential buyers are quite sophisticated when it comes to costs per square foot. If a speculative builder is considering a particular plan whose estimated cost per square foot seems too high, a good record of actual costs for the various major categories of similar homes can help to identify where the costs may be excessive.

Improve Efficiency and Quality—Many builders, remodelers, and developers see an improvement in both the quality of the work and the efficiency of their operations when they use cost control measures. If you discover that the cost of a certain major aspect of the work is running below the budgeted amount, you may find that the materials being used are of a poorer quality than specified and estimated. If the cost is running too high, the work may have been done wrong and torn out, or the plans may have errors in them. Either of these two problems could cause a decrease in the quality and efficiency of the project.

Improve Employee Morale and Loyalty—When a builder, remodeler, or developer is exercising sound cost control measures, an overall attitude of order and professionalism runs throughout the company's organization. This attitude often leads to improved employee morale and loyalty.

Increase Productivity—A company's attitude toward cost control systems may impact productivity, which is sensitive to overall job satisfaction. Construction workers usually take a great deal of pride in their finished product. When construction workers sense that the project is flowing smoothly and that all the other participants are doing their jobs in a proper manner, productivity tends to increase. Improving productivity and quality also decreases waste. A positive attitude from employees enhances the company's reputation and creates higher profits. On the other hand, anything that tends to disrupt the timely completion of the work or adversely affects its quality usually has a debilitating effect on productivity. Perhaps the most important aspect of cost control to most builders, remodelers, and developers is the sense of well-being that comes with knowing exactly where they are financially at all times. Why wander blindly through a mine field when you can make the journey with your eyes open?

Reduce Waste—A company that is conscious of its costs is more likely to be less wasteful than one that does not. Waste reduction is like safety in that it reflects an attitude more than anything else. The individuals who can impact waste reduction are more likely to make it a habit when cost control measures are integrated into the company's procedures.

Factors Affecting Record Keeping

Each factor that affects record keeping has its own special accounting problems. To a large degree the nature of the job usually dictates the accounting requirements. For example, if you are building, remodeling, or developing under a cost-plus contract, you must keep your records in such a manner as to support your billings. However a builder doing fixed-price contracts or speculative work has other needs. Both fixed-price and cost-plus builders need a system that tells them immediately if a particular aspect of the job needs attention.

Other factors may impact how you keep cost records such as—

- job location
- how materials and labor are handled
- use of equipment
- use of subcontractors versus in-house tradespeople

Job Location

Most builders, remodelers, and developers confine their operations to a relatively small geographical area. If they take on a project located some distance from their own home base, new problems may occur that will affect cost control records. New costs such as travel expenses, parking fees, and possibly different costs for worker's compensation and other insurance coverages can affect not only the profit but also how you keep records.

Although most small- to moderate-volume builders, remodelers, and developers tend to keep the bulk of their jobs in a fairly small geographic area, large-volume, well-financed companies can afford to go further afield to take on more jobs. However, even for them, job location can create cost control problems in addition to other management problems, such as the need to locate new suppliers and subcontractors and to obtain adequate parking without exorbitant fees.

The simpler and smaller the job, often the easier you can find capable subcontractors and labor and the necessary materials. As the job becomes more complicated, the more difficult the tracking and purchasing systems become.

Materials and Labor

Schedule Deliveries

For large, more complicated projects, the scheduling of deliveries also becomes complex, especially for hard-to-get or long-lead-time items. If these items cannot be incorporated into the job immediately upon delivery, protecting them from the weather, pilfering, and damage from normal activities creates new management and record-keeping problems.

Handle Materials and Labor

How you handle materials and labor affects how you keep records and institute cost control measures. Builders and remodelers face a different sort of problem regarding materials and labor from those faced by many developers. Builders and remodelers typically subcontract a large majority of their work. The single largest variable in construction is in the use of labor and equipment. Therefore labor and equipment also carry the potential for the greatest risks. In this context, labor is defined as workers who are your direct employees and who work by the hour.

Labor and equipment costs represent the greatest risks because productivity rates—how many units of something a worker can install or do in a given time period—may vary greatly. Many factors impact productivity, including those listed in Figure 1-1.

Increase Productivity

Worker productivity is an industry-wide problem. To achieve the best possible productivity, you obviously have to track labor costs so that you know how much work is being done for a given amount of labor on a daily basis. If you discover that the productivity is falling behind what you anticipated, you need to find out why. Many times you can cure the problem, but often you cannot. In many situations someone else causes the losses. If you give that person notice, the problem can often be solved. In these situations, if the person responsible for the decline in productivity cannot or will not take corrective actions, you may need to notify the person in writing for you to have a valid complaint and protect your rights. In any event, to the extent that you have direct labor working by

Figure 1-1. Factors that Impact Productivity

- weather
- adequacy of the plans and specifications
- the amount of overtime recently done by the worker
- the degree to which a job is overstaffed
- the degree to which other trades are working in the area
- worker morale and attitude, illness, and tardiness
- disruptions in the job rhythm or work sequence
- the learning curve associated with repetitive tasks
- owner occupancy
- access to the site
- delays in the work
- excessive changes
- acceleration of the work
- suitability of tools and equipment to jobsite conditions or size
- the fragility of materials, their ease of handling, and need for protection
- the worker's experience and competence
- availability of materials and tools

the hour, proper cost control measures can help to identify and often solve problems.

Purchase Materials

The method used in purchasing materials will influence how you use cost control measures in tracking their costs. (Chapter 5 covers many details regarding purchasing and materials management.) You will need to adopt a cost control system that is compatible with the way you control purchasing.

Use of Equipment

The use of equipment in a construction operation creates a whole set of unique problems and decisions for builders, remodelers, and developers with regard to the cost control system. Because of the current trend toward subcontracting the vast majority of the work, builders, remodelers, and developers today do not need as much equipment as their fathers and grandfathers needed. Today the subcontractor has the backhoe, the ditching machine, and the compressors. However builders, remodelers, and developers are often faced with the question of whether to buy or rent equipment. Of course the first question to consider is whether or not you have or can hire an employee who can operate the equipment since renting or owning typically means that you furnish the operator.

If you have no employee who can operate the equipment, you are probably better off engaging a subcontractor who can do the work.

On the other hand, developers may find the cost effectiveness of equipment ownership attractive. Owning large equipment almost always requires a lot of capital or good banking relationships. If a builder or remodeler goes to an equipment rental company to lease or rent a piece of equipment, he or she should expect the rental company to make a fair profit. If you rent a particular piece of equipment only once in a great while, probably purchasing it would not be profitable. But as your need for a piece of equipment increases, and as you use it more and more often, the decision of whether or not to purchase becomes more difficult. You could look at the question this way.

If you buy a piece of equipment, you are in the equipment rental business. The part of your company that owns the equipment rents it to the part that builds. The main reason to buy is to make the profit that you would be paying to a rental company. If you can purchase the equipment and maintain it for less than the rental cost, you should consider buying. Maintaining it includes all costs associated with the equipment (storage, fuel, repairs, and the like.)

You should use record-keeping and cost control measures to constantly track the costs so that you can determine how much to charge each job, how much must be absorbed by home office overhead, and at what point disposing of a certain piece of equipment is more economical than to keep it in good repair.

Use of Subcontractors

Within the past few decades, residential construction has evolved from doing most of the work with a builder's, remodeler's, or developer's own forces to subcontracting most of the work. Today's builder or remodeler is more likely to be a manager of subcontractors rather than a supervisor of workers. This trend evolved for several reasons. Many subcontractors are small-volume companies in which the individual workers are often related or well acquainted with one another. Good productivity is easily achieved because the supervisors often have a vested interest in the profits. Subcontracting work reduces some of the builder's, remodeler's, and developer's responsibilities. On a fixed-price subcontract, whether it is lump sum or unit price, the subcontractor is responsible for maintaining productivity. The subcontractors may supply some or all of the materials needed to do the work specified. The builder's, remodeler's, and developer's concerns are primarily reduced to quality and timeliness of completion.

Subcontractors who do the same tasks every day usually become more proficient at their work than someone who performs the work only occasionally. This specialization usually means greater quality and productivity, but you should not take it for granted.

Therefore your supervision role is more concerned with quality and less concerned with productivity. If a builder, remodeler, or developer is doing most of the work with direct labor, he or she experiences more difficulty expanding and contracting the volume of work being done as the demand increases or decreases.

Problems with cost control relating to dealing with subcontractors are different from those involving labor or materials. If a subcontractor does a small portion of the total job quickly, the cost control task is easy. The subcontractor will submit his or her bill, the builder, remodeler, or developer or an employee will verify the work, collect the insurance certificates, and pay and post the invoice. However large subcontracts often require partial payments. If handled properly, these payments can also be an easy task if you establish, up front, a time to inspect the work before you accept the invoice and a time for the subcontractor to present the invoice.

Backcharges are one of the biggest challenges associated with dealing with subcontractors, especially from a record-keeping and cost control standpoint. If a subcontractor damages the work of others to the extent that extra costs are incurred, the offending subcontractor normally is backcharged for those costs. For example, if the trim carpenter drives a nail through a water supply line in the wall, the plumber and drywall subcontractors may bill for the extra work. Accounting for a mystery offender can also be a problem. Often a builder or remodeler does not know who dropped something heavy and chipped the porcelain tub or broke the expensive plate glass window. Unfortunately sometimes workers do not report the damage they do, and the builder, remodeler, or developer must absorb unexpected extra costs.

Office Management

Many aspects of office management are enhanced by good cost control techniques. Subcontractors' invoices can be easily checked to prevent overpayment and to insure timely payments. Payroll records can be easily translated to production rates for future estimates. Change orders become easier to track. Invoicing customers or the bank can be done with greater accuracy, especially on cost-plus projects. When delays arise, the proper use of cost control techniques can help to minimize their harmful effects. Suppliers' invoices are more easily checked for errors, and customer service functions are facilitated.

Overhead

Tracking costs involves two kinds of overhead, job overhead and general overhead, and each is accounted for differently. Job overhead comprises all those costs that can be allocated directly to a particular project but represent charges for items that are not incorporated into the work. These costs are discussed and itemized in Chapter 2 under the heading of Job Overhead. However, for purposes of this discussion, job overhead includes such charges as the superintendent's time, permits, jobsite utilities, and sales costs. Many builders find that estimating and accounting for job overhead expenses are among the most difficult tasks they have to do. The way you set up your procedures for tracking these costs will depend largely on your own individual operation.

General overhead presents an entirely different set of challenges. Some builders refer to general overhead as home office overhead. These costs represent expenses for the office where your main business is housed. General overhead expenses are all those costs that cannot be, or cannot easily be, charged directly to a particular job. The way that a builder, remodeler, or developer sets up his or her general overhead expenses greatly affects how cost control records are kept. Among the many tasks performed by home office personnel are marketing and selling, processing change orders, preparing payrolls, invoicing customers, paying subcontractors and suppliers, and handling customer service. The degree to which you decide to break down costs incurred at the home office to the job level depends on your particular needs. A detailed analysis of these problems appears in Chapter 4.

External Factors Affecting Cost Control

Builders, remodelers, and developers keep cost records for two reasons—external and internal. The external reasons include filing taxes, securing financing, dealing with insurance companies, and securing any required bonds. The internal reasons include monitoring costs and creating historical cost records for reference purposes.

Manage Taxes

In trying to manage your tax liabilities, your cost records are essential. The payment of some taxes (such as withholding, unemployment, and sales taxes) is based on simple formulas, and little flexibility is possible. For these taxes, your records should provide for the easy retrieval of the necessary data. However for federal, state, and local income taxes, many options exist under the various tax codes to legally minimize taxes. A properly designed accounting system can help to provide strategies for keeping these tax burdens down.

Manage Financing

You also need to keep records in a systematic manner to manage your financing. You may need to borrow money to finance your operation, especially if you are primarily a production or speculative builder or a developer and to a lesser extent if you are a contract builder or remodeler. In these cases your financial needs will vary depending upon the terms and conditions of your contract.

For example, if you have a speculative job, you will need the wherewithal to build the project completely and sustain interest payments for a reasonable period of time. Contract houses and remodeling jobs can be either fixed-price or cost-plus contracts. In either of these two cases, you may need to meet interim payroll obligations or make payments to subcontractors or suppliers before you receive your progress payment from the owner.

You also will need to make capital expenditures periodically for equipment and other kinds of company property. If any of this financing comes from borrowed funds, you will need good records for your lender. Lenders like to deal with customers whose accounting and record-keeping systems are well thought out and well kept.

Whether you are borrowing money for a speculative project, for a line of credit to help meet daily cash flow needs, or to purchase a new vehicle, good record keeping plays a crucial role in securing financing. Lastly, sound financial management can help formulate strategies necessary for minimizing the cost of financing. These strategies include making accurate forecasts about revenues and planning your expenditures accordingly.

Manage Risks

Volumes have been written on risk management so this discussion is limited to its relation to cost control. Risk management is a comprehensive approach to measuring and handling exposure to losses. The three steps to risk management are to identify the potential risks that you face, determine your exposure relative to those risks, and decide how to handle each risk. The potential risks that a builder, remodeler, or developer needs to consider include—

- property losses on projects such as those covered by builder's risk insurance
- liability for worker's compensation
- liability for losses to nonemployees such as those covered by public liability and property damage insurance
- losses to the builder's own property
- motor vehicle damages
- various business losses such as business interruption, loss of a key person, or major medical losses
- inability to complete the project on time because of bad weather or unavailability of labor (mostly subcontractors) or materials, thus increasing financing expenses and possible penalties.

Properly kept records will help to prove the cause of and determine the amount of losses where such losses can be recovered from insurance or third parties. They also are a necessary part of the worker's compensation insurance process. As most builders, remodelers, and developers know, you will be responsible for paying your worker's compensation insurance premiums based on your records (which must include certificates of insurance from your subcontractors).

Bonding

The last external reason that some builders may need to keep good cost records has to do with bonding requirements. A builder, remodeler, or developer might have to obtain several kinds of bonds, but the two that depend most upon good records are performance bonds and payment bonds.

A performance bond is a contract between the builder or remodeler and the bonding company or surety that guaranties the owner that the contract will be performed and that the bonding company will make sure the home or the remodeling job is completed in substantial accordance with the terms of the contract between the owner and the builder or remodeler. If the builder or remodeler defaults and the bonding company has to finish the job, the builder or remodeler may be liable to the bonding company for its losses.

A payment bond guaranties payment for labor and materials used on the project. The primary purpose of this bond is to protect the owner against liens. Performance and payment bonds are common in nonresidential building construction and are typically required for publicly owned projects. Home builders and remodelers seldom have to produce bonds of this nature.

In the event that you are required to furnish a performance or payment bond, the surety will investigate your experience, financial status, and professional abilities. To be bonded, you may need to provide information about your cost control and accounting systems to the surety.

Internal Factors Affecting Cost Control

While all the external reasons mentioned above are important to a homebuilding, remodeling, or land development business, most businesses keep records for what are termed internal reasons. These reasons all revolve around assessing and improving profitability. Builders, remodelers, and land developers have many intangible reasons to be in the building business. Most people in construction take a great deal of pride in their finished work. They get a special feeling from standing back and looking at a product that they have personally created.

All these kinds of reasons notwithstanding, a primary reason to be engaged in the building business is profits. Proper cost records are essential in assessing the profitability of a homebuilding, remodeling, or land development company and to determine which costs need tighter control to increase profit for the owners of the company and its investors (if any). From well-conceived and well-kept cost records you can easily determine—

- your profit or loss
- what dividends are in order
- the degree of your company's liquidity
- your cash flow position

Monitor Costs

Of the two main internal reasons to exercise cost control, the first is to enable the builder, remodeler, and developer to monitor their ongoing costs. Making intelligent decisions requires that you know whether or not you are making or losing money not only at the job level but also on an overall basis.

If you are going to make a trip across the country by automobile, you can just get in your car and go, and when you get there, you are there. However, if you need to be there by a certain time, the first thing you should do is to determine how long the trip will take. Then you should establish your rate of travel and determine when you need to be at certain intermediate points to stay on schedule. If you fall a little behind, you can make whatever adjustments are necessary to get back on schedule.

The way you exercise cost control on a job is somewhat similar. The first thing you do is to determine how much it should cost. In doing so, you also determine how much each phase of the work should cost. When you start building, remodeling, or developing land, if you keep track of how much you have spent on each particular phase and compare that to your budget for that phase, you can tell whether you are ahead or behind.

For example, you may have a budget for rough grading of $6,000. If you have expended $4,000 and are less than half finished, you need to find out why. Of course, as each phase of the construction finishes, you can compare your costs to the budget to see if the total project costs are going to be as expected. This step can be especially important for speculative jobs because you do not want to sell a project below cost.

Keeping cost records for each phase of the work and comparing them to the budget can help to correct mistakes and prevent future problems. For example, if material costs are running high, you may have a problem with your estimating. On jobs in progress, speculative builders may be able to correct the problem or adjust markup accordingly. On future work you may be able to avoid the problem altogether. For example, assume that you are starting five new homes, one right after the other. Further assume that your rough-grading subcontractor is working for an hourly rate. If the budget for the first job is substantially exceeded, you may have a subcontractor who cannot work efficiently. If this situation is the case, you will be a lot better off knowing this early rather than when the jobs are all done.

Decision Making

The essence of good management is the ability to make correct decisions. Timely reports and a reliable cost control system give builders, remodelers, and developers more options in making decisions. Applying history (what you learn from past jobs) helps you make correct decisions on new work.

Cash Flow

An important element in the monitoring of on-going work is the proper management of your cash flow. Your cost control records can be an invaluable asset in projecting cash-flow requirements and helping you to meet those related needs.

Historical Database

The second main internal reason to institute a cost control system is to develop historical records. A builder, remodeler, or developer who keeps good records can more accurately estimate the cost of new work and more accurately determine the correct price for changes in the work. Also, when a builder, remodeler, or developer is reviewing an estimate, he or she can more easily spot prices that are out of line. Different reasons exist for variances between the estimated cost and the actual cost, such as

(a) mistakes in the estimate and (b) designs that are not cost effective and cause excessive scrap and waste in the use of materials.

Remodelers and especially restorers and renovators often develop their own price books because they use so many different kinds of materials that may not be readily attainable at the local lumber company.

A good cost control system will enable the builder, remodeler, or developer to quickly check unit prices of different parts of the work and compare them to a historical statistical average to aid in bidding and negotiating new work. These records also will help in doing cost-plus work for a variety of reasons. Most builders and remodelers who do a lot of cost-plus work need to have a reputation of doing good work. This reputation is created and enhanced not only by the quality of the work itself but also by the professional way the builder or remodeler handles the job. When you are negotiating a cost-plus job, if you have good records on costs of various phases of the work, such as the average cost per bathroom or the average cost per square foot for heating, ventilation, and air-conditioning, you can more easily suggest alternatives that will help your customer. Along these same lines, on any kind of project you have, whether it is a custom or speculative job, your records will help in evaluating alternatives as is done in a formal value engineering procedure.

Your historical cost records will also allow you to determine whether to do a certain aspect of the work with your own hourly forces or with subcontractor forces. For example, you may find that certain subcontractors whose labor need not be particularly skilled are making an inordinately large profit. In these cases you may choose to do that work with your own forces or even put a new subcontractor into business and share in its profits. Sometimes a builder, remodeler, or developer is faced with a decision of whether or not to go into a specialized line of work such as siding or roofing. These choices are often made easier if you have good cost records. If you do work as a speculative builder or as a developer, your choice of materials becomes one of the more important considerations. The installation costs for many materials differ from the installation costs for acceptable alternatives. Good cost records are helpful for these kinds of management decisions.

Ways to Keep Records

Of course, cost records may be kept in various ways. Some builders, remodelers, and developers might keep records for external purposes differently than they do those for internal purposes. The aim of this book is to focus on records that are beneficial for internal use, that is, to help monitor costs and establish cost records to help estimate new work.

As explained in greater detail in Chapter 2, all costs may be divided into two categories—direct and indirect costs. Direct costs are those that can be readily charged to a particular job while indirect costs are those that cannot. Records for direct job costs may be kept simplistically, or they may be kept in a highly sophisticated manner. More detail and sophistication does not necessarily make your job cost records more beneficial. You will need to pick a level of detail and try it. If you find that everything you are recording is helpful, you might try a little more detail. If you find that your records are not totally helpful, you can try to simplify them.

Total Job

The approach developed in this book will allow builders, remodelers, and developers to keep cost records in various degrees of detail from the most rudimentary to the highly sophisticated. The simplest approach is to keep a running total cost for each job. When the job is over, you will know what it cost. The next higher level is to divide the cost for a job into categories and compare the cost of a particular work item to the estimated or budgeted cost. Regardless of the method you use, keeping records of your costs can and will make you a better business person.

Divisions by Category

The following chapters outline in detail how builders, remodelers, and developers may subdivide the various costs associated with their jobs into different categories. Examples show one way this breakdown of costs may be done. Your own needs may require more or less detail. Additionally after you have tried one way to categorize the costs, you will probably want to modify your method to better suit your needs. However, within a given job cost category, a division of the costs into materials labor, equipment, and subcontracts gives the best results.

The builder, remodeler, or developer must collect and analyze the information contained in all these records before he or she makes the crucial decisions about a new job. If the system cannot do that, it is crippled as a management tool. Therefore the cost control system should collect the right information as defined by the needs of the business and provide both regular and on-demand reports to the builder, remodeler, or developer in time for decision making. Listed below are items to remember in keeping records:

- All costs are charged directly to a job or to general overhead.
- General overhead expenses must be absorbed by the jobs on a percentage basis.
- Profit is what is left after paying all job expenses and general overhead expenses.
- The only businesslike way to make proper decisions about your business is to know as soon as possible whether your jobs are ahead of or behind your budget.

Two

Plan Your Approach to Cost Control

The construction business differs from manufacturing businesses in several respects. Building, remodeling, and development projects are almost always unique efforts that utilize craftspeople who may not necessarily have worked together in the past. They also may not have built that exact house previously and almost certainly will not have done an exactly similar remodeling project or developed an exactly similar tract before. For most of these jobs, you have to bring the factory to the jobsite each day. In addition, except for some remodeling projects and modular houses, much of the work is outdoors and subject to changing elements. In this setting, problems related to cost can quickly get out of hand. Builders, remodelers, and developers must use proven management techniques to ensure high-quality production and profitable results.

Job Costs

The main basis of a builder's, remodeler's, or developer's cost control system must focus on determining and analyzing expenses and income from each individual project. In other words each job is a separate profit center for which you can determine the total costs to date at any time. As a job is progressing, you should know whether it is making or losing money, and when the job is finished, you should know how much profit or loss you had on that job.

To start a cost accounting system, you must establish a framework that (a) suits the needs of your particular business and (b) enables you to establish budgets and post ledgers accordingly. Every expense your business has must be charged either to a job account or to the general overhead account.

To start job cost accounting, you must establish a framework that (a) suits the needs of your particular business and (b) enables you to establish budgets and post ledgers accordingly.

The most rudimentary system for implementing cost control procedures would be to charge job costs and expenses to a job without further breakdown. To keep track of these costs you could set up a notebook with each job having its own page. At the top of each page you could write the name of the job and the total budget amount for the job. As you incur costs for a job, you could write down the amount on the pages set aside for that job.

If you keep the entries up to date, at any time during the progress of the work, you could determine the total direct costs for that job to date. When you complete the job, you would know the total cost of the work including job overhead costs and be able to compare it to the budgeted amount. Likewise you could list general overhead expenses on a general overhead page as they occur. At the end of the fiscal year, you would know how much your general overhead was for the year.

This simplistic system would provide several benefits. First, you would know if your business is profitable or not. Second, you would know which jobs made money and which jobs lost money. This knowledge would enable you to review your efforts and your estimate to see where any problems were. Of course, this process would not help a job that had already lost money, but it could help you to better plan and estimate future jobs if you could identify where and how an error occurred. (A major goal of cost control is to plan better.)

When a builder, remodeler, or developer estimates the cost of a new project, he or she breaks the project down into several categories. The detailed example given in *Estimating for Home Builders*[3] divides a new home project into 17 categories plus the cost of the land which brings the total to 18. Each category is estimated independently, and the cost of each individual category is totaled to give the cost of the whole job.

The categories (discussed below) are listed roughly in the same order as a job is built. For example, one of the early categories is site work, which includes such items as clearing, fill material, demolition, and erosion control. This particular portion of the job is usually completely finished early in the course of the project. The examples for builders used later in this book are based on this 18-category breakdown.

Land

The price of the land on which you build may or may not be a cost to you. If you buy the land and your sales or contract price includes the property, you should include the cost of the land and all the improvements you paid for in the cost figures for that project. For example, a speculative builder will almost always purchase the land and include its cost in the sales price. Occasionally a contract builder will purchase the land in conjunction with his or her contract with the purchaser of the house to be built. Builders buy the lot in these situations to protect themselves. If the home buyer becomes guilty of a major breach of contract, the builder is in a better position legally and financially if he or she owns the land rather than if the home buyer owns it.

Costs to be allocated to the category for land may include all the costs associated with providing utilities to the property. For many residential lots, the price includes water and sewerage because they are developed lots. For other lots you may have to drill a well, install a septic system, or pay for offsite utility development (such as running utilities in from the main road to the lot).

Some builders charge these costs to the land category so they can compare the price of the land to a comparable piece of land for which all utilities are furnished. Simply adding the cost of land and the development costs allows for the same comparisons.

Job Overhead

This category of job overhead includes all the costs associated with a particular job that are not reflected in the materials, labor, equipment, or subcontracts that go into the actual building of the job. The direct job costs in this category may be broken down into several subcategories. These subcategories are listed in the appendix and are discussed in detail below.

Architectural and Engineering—Any professional services required for the job such as plans and specifications or surveying are accounted for here.

Land Closing Costs—If the builder or developer purchases the property, he or she will pay closing costs on the purchase of that land. Those closing costs could include such items as—

- appraisal fee
- discount points or origination fee
- title examination
- attorney's documentation fee
- recording fees
- other attorney fees
- photographs
- taxes

Construction Loan Closing Costs—A builder or developer may borrow the construction money to build a speculative project. Costs associated with a construction loan are part of job overhead including—

- appraisal fee
- discount points or origination fee
- title examination
- document preparation fee
- recording fee
- other attorney fees

Permits and Fees—In most jurisdictions a builder, remodeler or developer must purchase a building permit prior to commencement of a construction project. Costs associated with building permits, homeowner's association assessments, warranty fees, and inspection fees fall under this section of job costs.

Tap Fees, Utilities—In many cases you might have to pay fees to connect new construction to existing utilities such as tap fees for water or sewer service. In addition to these fees, other utility costs such as temporary power hook up costs, temporary utility expenses, and water usage costs are accounted for here.

Insurance—Typical insurance costs would include premiums for insurance that the builder, remodeler, or developer can allocate directly to the job such as builder's risk, liability, and completed operations insurance. Worker's compensation insurance should be charged directly to the category for which the work is being performed.

Interest—If a builder, remodeler, or developer borrows money to finance any part of a project, the interest that he or she incurs needs to be charged to the job as a direct job cost.

Sales Costs—If the job is a speculative project, such as a speculative house or subdivision development, usually costs are associated with the sale of that project. These costs may include—

- appraisal fees
- discount points or origination fee
- title examination
- document preparation
- recording fee
- attorney's fees
- photographer
- taxes
- sales commission

Field Overhead Expenses—You would include the cost of the superintendent in this category. The cost may be spread over the several jobs that a superintendent is running. Other costs in this category would include temporary toilets, temporary fences, job site offices, job site storage facilities, and guards, if applicable.

Site Work

The cost of all work necessary to clear the site and prepare it for construction fall under this category. Such costs include—

- the tasks of clearing the building site
- any hauling and fill materials needed to prepare the site
- rough grading
- demolition and removal of any existing structures
- installing any necessary utilities
- erosion control

- compaction
- weather-tightening of portions of the house opened up by remodeling activities
- Labor to accomplish the work listed in this category

Footings, Slabs, and Walls

This category includes all costs associated with work on the concrete footings, slabs, and walls that are necessary for the construction of the houses or buildings such as—

- batter boards
- concrete formwork (including form boards, brickledge, stakes and braces, and special forms if needed)
- excavating footings (Some builders combine forming and footings into one category.)
- reinforcing steel (dowels, anchor bolts, rebars, wire mesh)
- porous fill material spread under slabs
- termite treatment of the soil and termite shields
- vapor barriers to go under the slabs
- concrete
- concrete placement and finishing
- curing costs

Masonry

This category includes all the materials and labor associated with a masonry fireplace. Even if the fireplace is prefabricated with prefabricated chimney sections, it goes in this category. Placing prefabricated fireplace units here makes comparison of costs with masonry fireplaces easier. The items included under masonry are as follows—

- masonry units (brick, block, stone, concrete masonry units, and any special masonry
- fireplace (prefabricated units with chimneys, firebrick, dampers, steel angles, flueliners, and common brick)
- mortar and all the associated materials (mortar mix or Portland cement, masonry sand, color or mortar dye, admixtures that might be required, and lime)
- metals used in masonry construction (wall ties, metal lintels to go over openings, all metal wall reinforcement)
- waterproofing materials to include drainage materials (such as gravel or geotextile fabric, piping to provide drainage from waterproofed areas, and waterproofing and drainage labor)
- stucco or synthetic stucco
- scaffolding used to install the masonry
- masonry labor

Framing

This category is subdivided into nine sections that cover all the materials and labor necessary to complete the framing and exterior millwork.

Floors—This subcategory includes sill plates that support wooden floors, floor trusses, beams, joist hangers or ledgers, joists, bridging, sleepers for floors on concrete, subflooring, and any nonmasonry piers.

Wall Framing—Materials necessary to frame the walls include treated plates, untreated plates, studs, headers, gable framing, blocking, corner bracing, sheathing, and posts and beams.

Roof Framing—This section includes all the structural components of the roof system such as trusses, ceiling joists, rafters, ridge material, hip and valley members, bracing and strongbacks, subfacia, lookouts, band materials, roof sheathing, plywood clips, and felt.

Exterior Siding and Trim—Materials include siding, facia, soffit, frieze, trim boards, molding, louvers, vents, all windows and skylights, exterior doors and locksets.

Stair Framing—Material to be installed in the framing stage fall under this subcategory such as landings, stringers, risers, treads, and banisters that are installed during the framing stage.

Exterior Decks—These items in the framing category would include posts, beams, joists, blocking, decking, rail materials, benches, stairs, and hangers.

Miscellaneous Carpentry—This group includes other items that carpenters install such as fences, skirting around porches, lattice, shutters, weatherstripping, screen enclosures, aluminum cladding of trim details, and hatchways.

Fasteners—This subdivision of framing involves a range of materials such as adhesives; nails, nuts, bolts, and screws; and column bases.

Framing Labor—The last section of the framing category is the labor to install all of the materials in the above subcategories. Any demolition labor required, such as on remodeling jobs, is accounted for here. This division includes subcontract labor, in-house labor, and all equipment expenses for items such as compressors, cranes, and scaffolding that are incurred by the builder or remodeler.

Roofing

This category covers all—

- materials associated with roofing such as roof edge, valley, and roofing material such as composition shingles, wood shakes, tile, metal, slate, and so forth

- all flashing
- nails and fasteners
- built-up or single-ply roofing
- ice and water shield
- ridge vents
- labor to install these materials and/or to rip off old material for reroofing

Plumbing

This category includes the basic plumbing package from the plumbing subcontractor. You would also list in this category—

- any one-piece vanity tops that are not a part of the basic plumbing bid
- any tubs or showers not in the plumber's contract in this category (such as whirlpool tubs or hot tubs)
- all shower and tub enclosures whether they are one-piece or marble units or ceramic tile (Keeping these items in this category allows you to make meaningful comparisons between one-piece tubs and showers and other types.)
- shower or tub doors
- sewer and water tie-in expenses (for lines that go from the house to the public sewer or septic system and to the public water system or well)

Electrical

This section covers the basic electrical package to be provided by the electrical subcontractor. If the electrician's contract does not provide for the following items, you should account for them here:

- lighting fixtures
- telephone rough-in
- television wiring
- sound system wiring
- intercommunication system
- burglar alarm
- burglar alarm wiring
- For a remodeling job bringing the existing electrical system up to code

Heating, Ventilation, and Air-Conditioning (HVAC)

The basic HVAC package is in this category. Any attic fans or other ventilating equipment not included in other subcontracts appear in this category. For example, the custom and practice in your area may be for the plumber to provide both the fans and the ductwork to vent bathrooms to the outside. If this custom prevails in your area, you may prefer not to break out the cost for this part of the work as a separate item unless one of the bids you are comparing does not include it.

Insulation

This category includes insulation for the attic or ceiling, the walls, slab edges or basement walls, rigid insulation for the roof, and any special insulation not already covered. Any labor to install insulation that is not a part of an insulation subcontract is accounted for here, including caulking, foaming, and chinking.

Drywall

Under this category, you would account for all drywall materials including regular, moisture-resistant, and fire-code drywall. Post here any costs for nails, screws, tape, or joist compound not covered by a subcontract. Also include labor to install the drywall and to finish or tape and float it.

Interior Trim

All the materials that a finish carpenter might install go into this category including—

- all moldings such as base, crown, casing, chair rail, jambs, cased openings, window trim, mantles, and special moldings
- paneling
- interior doors
- shelving
- closet and shower rods
- mirrors
- hardware (privacy and passage locks, dummy locksets, doorstops, bath accessories, disappearing ironing boards, dryer vents, and soffit vents)
- stair trim (finished risers, rails, threads, and skirts) and disappearing stair units
- All fasteners necessary to install the trim materials
- labor to install all the items in this category

Paint and Wallcovering

All paint and related materials and wall coverings that the builder or remodeler purchases fall under this category. In most parts of the country all materials are furnished by the painting contractor. You can account for all tile wall-coverings (such as ceramic tile) that are not a part of the tub enclosures or showers in this category. You would also charge the labor to do the paint (if it is paid for separately) or install the wallcovering to this category.

Floor Covering

All finished floor-covering products and the labor to install them belong in this category. The materials include carpet and pad, resilient floors, wood floors, hard floors such as tile, slate, and brick. If the builder or remodeler incurs any separate labor expenses associated with floor covering, it is posted here.

Cabinets

The various subdivisions of this category are kitchen, bathroom, and special cabinets; cabinet tops; backsplashes; cabinet hardware; and cabinet installation labor. You must decide the best way to account for the cost of these various items because the method you choose has to suit your business. For example, this method calls for one-piece vanity tops such as cultured marble tops to be posted under the plumbing category because they contain the sinks. If the vanity top is plastic-laminate glued to plywood with a separate vanity sink, you might charge the sink to plumbing and the top to cabinets. The key in setting up these subdivisions is to determine how you can best categorize the items to make estimating easier for you and to provide the best arrangement for making cost comparisons.

Appliances

The various subdivisions of the appliances category might include dishwasher, range or cooktop, exhaust system, oven, microwave oven, refrigerator, disposal, central vacuum system, and any other appliances. You will need to add to or subtract from this list depending on the needs of your business.

Exterior

In this category you account for the labor and materials for all the final outside work such as—

- final grading
- landscaping (fill material, topsoil, trees, shrubs, mulch, sod, seed, and fertilizer)
- walks, drives and patios (including formwork, forming labor, concrete, concrete labor, base material, spreading and compacting base material)
- concrete finishing labor
- brick, stone, and tile or other materials for walks and patios
- irrigation or sprinkler systems
- gutters, downspouts, splashblocks, and drainage systems for gutters
- ornamental ironwork used in landscaping
- fencing not accounted for in miscellaneous carpentry
- all cleaning including intermediate and final cleaning, inside and outside, supplies, trash removal, and hauling and the labor for doing it (Some builders and remodelers subcontract these tasks, and the subcontractor provides all the supplies.)

A complete listing of the breakdown given above appears in the appendix. This list is basically intended for new home builders, but remodelers can use it as a starting point to make their own list of categories for cost control. They may use some of the categories on nearly every job, and some they will use occasionally. Others they will not use at all. Remodelers may use categories that do not appear here depending upon how they choose to break down their work. Remodeler-only categories are discussed in Chapter 6. Categories used only in land development appear in Chapter 7.

General Overhead

As mentioned earlier, every expense you have must be charged to a job or to the general overhead account. Cost control requires a builder, remodeler, or developer to have a clear understanding of (a) what comprises general overhead costs and (b) how to track and evaluate such expenses.

Of course, if you get a bill for materials that went to two different jobs you would not charge it to general or administrative overhead. You would break it down and charge it to the individual jobs. General overhead is made up of all those general business expenses involved in running the company office. These expenses support the business's overall building effort and cannot easily be charged to a specific job. They include rent, utilities, advertising, marketing and sales, phones, beepers, taxes, interest, professional fee, training, education, conventions and seminars, salaries and commissions. (A complete listing begins with the subhead Salaries.)

You need to track your general overhead costs for a specific time period, usually a year, and determine what percent of your total volume of sales it represents. For example, assume that a builder's general overhead ran $80,000 for a particular year and the job costs for that year totaled $1,600,000. The general overhead needs to be allocated or spread over the various jobs in a proportional manner. If one job had cost $200,000, the builder would need to charge $10,000 of the general overhead expenses to that job because that job was one-eighth of the total volume and, therefore, must absorb one-eighth of the general overhead. Another way to look at general overhead allocation is to divide the general overhead by the volume to get the general overhead percent. For this company the general overhead percentage was 5 percent ($80,000 ÷ $1,600,000). Therefore this job had a total cost of $210,000, $200,000 in direct job costs and $10,000 in general overhead costs. If you sold the job for $240,000, theoretically you made a profit of $30,000.

Knowing what percentage of your costs the general overhead represents can help you in estimating the cost of future work so that you account for all your costs of running a professional business. The best cost control is done in planning a job (before anything becomes unchangeable). After you determine the direct job costs, you add your markup to that cost. The markup is composed of two parts, general overhead and profit. Assume that your general overhead runs 6 percent. If you use a markup of 16 percent, you will have a profit of 10 percent built into your job. If you do not include any profit in your markup and if you receive exactly the direct job costs and your general overhead, you will be able to pay all your bills for the jobs and for the company office, but you will not have any money left over. You will have no profit. (The paragraphs that follow suggest one way that the general overhead expenses may be broken down into categories.)

Salaries

You should charge to general overhead the salaries of everyone who spends either part time or full time in the business and whose efforts cannot be charged directly to a job (including your salary or draw). Some small-volume builders, remodelers, and developers do not consider their own pay as a general overhead expense. Instead they pay themselves out of the profits, which are elusive. Industry experts agree that the best way to handle this situation is to pay yourself a reasonable salary and charge it to general overhead. By following this procedure, you will be able to more accurately analyze the exact nature of your profits or losses. Of course, if the total income for your company is less than your direct job expenses plus your general overhead, the first place that gets cut back is your own salary. If you can see a shortfall coming, you reduce expenses or increase production and sales and thereby cut your losses. Hopefully the exercise of cost control measures will prevent shortfalls.

The salary category typically accounts for business owners who actively work for the company as well as office managers, sales managers not on commission, estimators, administrative assistants, bookkeepers, secretaries, production managers, and other clerical workers. Of course, if any of these people spend time working on a particular job in some capacity, such as a superintendent, you should charge that part of their pay to the job.

Office Labor Burden

All payroll taxes for main office employees belong in this category along with worker's compensation, insurance premiums and health benefits; pensions; vacation and holiday pay; sick, personal, and emergency leave pay.

Office Expenses

Expenses associated with occupancy of the main office such as rent, office equipment and furniture depreciation, repairs, maintenance, utilities, and supplies are included. Sometimes long-distance telephone calls made from the office pertain to a particular job. Some builders, remodelers, and developers use phone services that require an access code that separates long distance charges to various accounts. Other builders include all long-distance charges in general overhead.

Computer Expenses

This category accounts for all costs associated with owning, maintaining, and operating computers including software, hardware (equipment), and supplies.

Vehicle, Travel, and Entertainment

Depreciation on company vehicles, lease payments, all vehicular expenses, travel expenses not associated with a particular job, vehicle or gas allowances, and entertainment expenses are included in this category.

Taxes and Licenses

This category accounts for sales taxes for items not connected with a job, real estate taxes not associated with projects, and all licenses.

Insurance

You would list in this category insurance costs not chargeable to any project or other category, for instance, fire and extended coverage on office buildings and general liability insurance.

Professional Services

This category includes fees paid to outside accountants, legal fees, and to marketing consultants. It also would include any consulting fees not attributed to a particular project. You should charge any professional fee that benefits a particular project, such as surveying, as a cost against that project. Travel and registration expenses for seminars, training, and conventions are charged to this category.

Depreciation

Even though your firm may use large, expensive equipment on various projects during a given year, the depreciation and other fixed expenses must be paid whether the equipment is used or not. Since these are recurring expenses, many builders and contractors charge them to general overhead. Depreciation on buildings, furniture, and office equipment is charged here as well as depreciation on company-owned vehicles.

Miscellaneous General and Administrative

This category would account for any other general overhead expenses that cannot be charged to one of the above-listed categories. Such expenses would include such items as bad debts, contributions, dues and subscriptions, and small tools. A complete listing of the general overhead breakdown appears in sections 800-899 General and Administrative Expenses in *Accounting and Financial Management for Builders, Remodelers, and Developers.*[4]

In using these lists of cost control categories, you will need to group the various items as your needs dictate. However you should follow these basic rules in Figure 2-1.

Figure 2-1. Checklist for Planning Your Cost Control System

- The way you group your items into categories for cost control purposes must duplicate exactly how you group the same items in your estimate. This practice is necessary if you are to make meaningful comparisons.
- Try to group items together within a category so that the materials in a category are installed by a single subcontractor if possible.
- Set up the categories so that you can close them as soon as possible and thus better monitor your costs.
- Divide all the expenses that you encounter on your jobs into a manageable number of categories.
- Use your categories as a basis to estimate the costs of your jobs.
- Charge every direct job cost to the job.
- Determine the best way to subdivide your own indirect or general overhead expenses.
- Refine your system as time goes by to better utilize your information.

Three

Track and Analyze Job Costs

Builders, remodelers, and developers have two methods of accounting for costs: the cash method and the accrual method. In the cash method, when you write a check to pay a bill, you post it to the appropriate ledger. In the accrual method you post the ledger when the invoice is received and approved for payment. When you use the accrual method your cost reports are more nearly up to date than when you use the cash method, especially if you let your bills go 30 to 60 days. The cash method is easier to do, but it lacks the capability to be a timely tool for analyzing costs.

Set Up Cost Categories

The object of tracking job costs is to know if you are above or below your budgeted amount in time to do something about it. Your cost control system can tell you what profit you have made on any job and create a record to use in future estimating and planning. Chapter 2 discusses one way in which you may subdivide the costs of a typical new residence. Later chapters consider remodeling and development projects. You should develop your own system that best suits your needs. In any event your categories should combine related items, and every cost item should fit into a category.

A rule of thumb for cost control is to avoid putting a miscellaneous category in your job cost ledgers. If you do, too many items that should be accounted for elsewhere end up in miscellaneous. Not accounting for your costs by placing them in inappropriate categories completely defeats the goals and objectives of cost control. The general overhead ledger includes a miscellaneous category, but that ledger is used differently from job cost ledgers.

After you set up your categories and the various items in each category, you will probably make a few changes as time passes to reflect experience or changes in the kind of work you are doing. The breakdown of job costs serves as a method of accounting for those costs and as a checklist for estimating new work. Whatever method you use for listing the cost items for the estimate, you also must use it for the cost control system to keep your comparisons valid.

You can design and use job ledgers and a general overhead ledger yourself and keep job costing completely in-house. You will be using the ledger to help manage the job. Managing the job—as opposed to reporting for taxes—requires that your job ledger track information in a way that allows you to make management decisions concerning the future. The records you need for external purposes such as for taxes and for lenders tend to report on events in the past.

Establish a Budget

The first step in performing job cost control is to do a detailed estimate. The next step is to establish the budget for each category based on the estimate. (The budget acts as your cost goal.) As an example, the estimate for a house with a living area of 2,472 square feet and a total area of 3,347 square feet is shown in Figure 3-2. Please note that the labor burden used for this example is 29 percent and the general overhead is 6.2 percent. Your labor burden and general overhead may be much higher depending on your operation. These percentages are for illustration purposes only. You must determine your own percentages. The budget for the 18 categories is summarized in Figure 3-3.

The figures in the column labeled Raw Costs come straight from the totals of each category of the estimate. However these numbers do not contain labor burden or sales tax; you must add these at the end. For example, look at the totals for the job overhead category. They contain no materials so no sales tax is added. However these totals do include $3,750 for direct labor to cover the superintendent's pay. The total cost to the company for the superintendent will be this amount plus the labor burden which includes payroll taxes, insurance, and fringe benefits. The labor burden is assumed to be 29 percent or $1,088. (Your labor burden may be quite different. The purpose of this exercise is not to recommend any particular percentage but to demonstrate a process.) Adding $1,088 to this category brings the budget for job overhead to $28,338.

In the footings and slabs category, $4,750 is estimated for material costs. Adding 7 percent for sales taxes on materials increases this category by $332 and brings the total budget for footings and slabs to $8,619.

Increasing each of the other categories for materials by 7 percent for sales taxes establishes the final budget. Of course, you will need to use the appropriate sales tax rate for your area as well as your appropriate labor burden.

The reason for adding the sales tax to each category is that you will be paying taxes when you pay for the materials on a category-by-category basis. Adding the tax to the budget makes sure you are comparing like quantities.

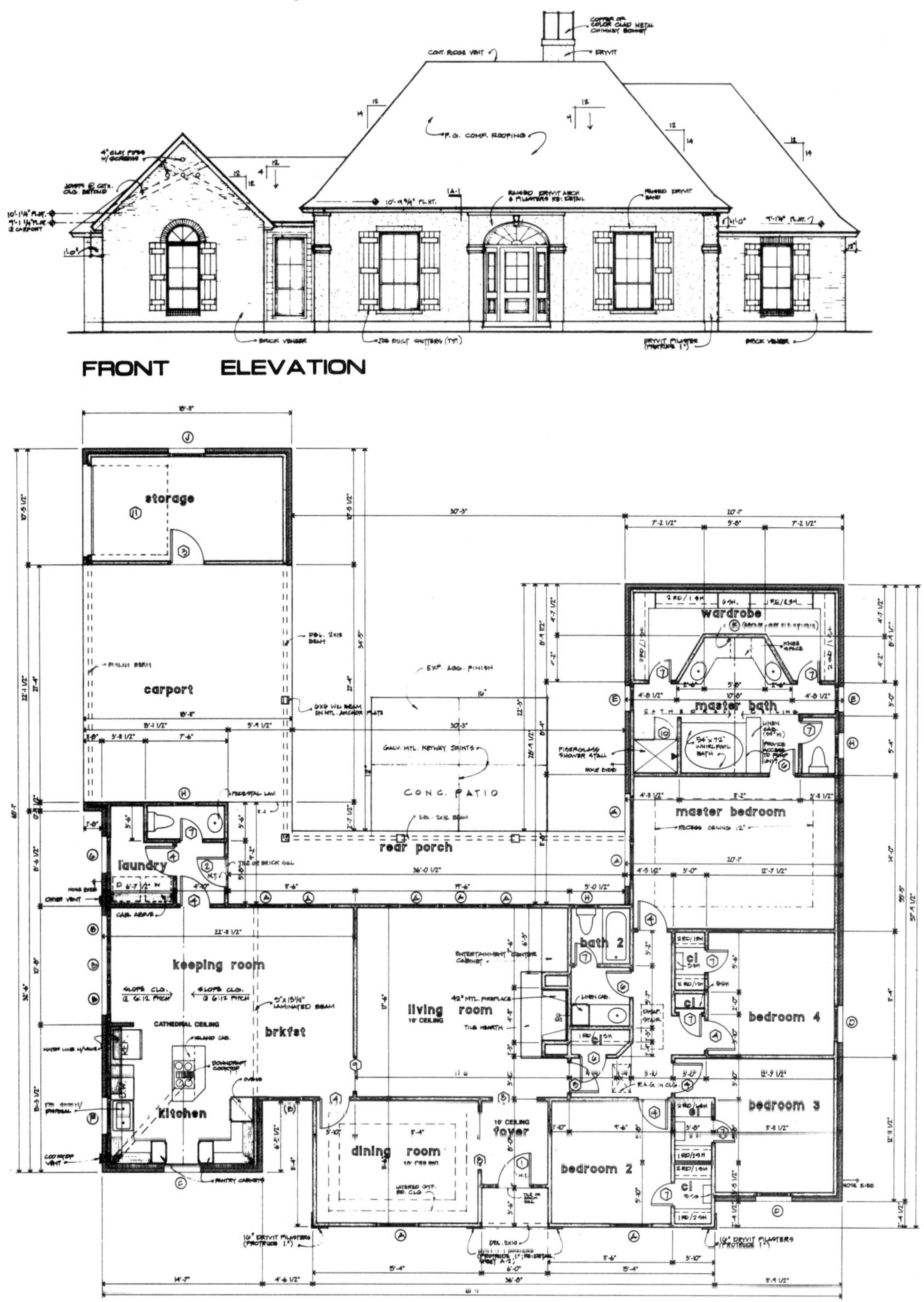

—Courtesy of David E. Blackwell, Baton Rouge Plan Service, Baton Rouge, Louisiana.

3-1. Sample Floor Plan and Elevation

This house floor plan and elevation provided the basis for the estimate and budget for the various cost control categories. Your houses, estimates, and budgets may have items that do not appear in this house and items in this one may not be included in yours. No matter what house is involved the principles of estimating, budgeting, and controlling costs remain the same.

ESTIMATE WORK SHEET

JOB Lot 175 Bally Acres P. 1

Description		Materials	Labor	Subcontract & Equipment	Total
Land Costs					
lot				57,000	
septic system				4,500	
				61,500	61,500
Job Overhead					
Plans				500	
Lot & const. loan closing					
appraisal				300	
discount pts				1500	
title exam				250	
document prep				50	
recording fee				85	
abstract				65	
Permit	$ 20 plan review plus $ 3/1000 cost			470	
Tap fees	Water - $250, Temp power hook up - $25			275	
Temp utilities				75	
Insurance					
bldr's risk				200	
completed opns				50	
gen liability				30	
Interest	300 (1+2+3+4+5)			4500	

MFD IN USA

FRANK R WALKER CO. PUBLISHERS, LISLE, ILLINOIS

Figure 3-2. Sample Estimate

ESTIMATE WORK SHEET

JOB Lot 175 Bally Acres — P. 2

Description		Materials	Labor	Subcontract & Equipment	Total
Sales commission	250,000 x .06			15,000	
Site expenses					
superintendent	45,000 /12 houses per year		3750		
toilet				150	
			3750	23,500	27,250
Site Work					
Lot grading	$ 200 mobilization 16 hr at $65			1240	
Erosion control				100	
				1340	1340
Footings & Slabs					
Batter boards	50 / 2x4x8 1.79	90			
Formboards	24 / 2x10x14 9.46	227			
Brickledge	14 / 2x6 x 14 6.65	93			
Stakes	200 0.20	40			
Braces	50 / 2x4x8 1.79	90			
Form labor	3347 ⌀ 0.50			1673	
Dig & spread	3347 ⌀ 0.25			837	
Rebars	50 #5's 3.99	200			
Wire mesh	5 rolls 34.60	173			
Porous fill	41 yds 8.00	328			
Termite treat	3347 ⌀ 0.15			502	

MFD IN U.S.A. FRANK R. WALKER CO., PUBLISHERS, LISLE, ILLINOIS

Figure 3-2. Sample Estimate (Continued)

ESTIMATE WORK SHEET

JOB Lot 175 Bally Acres　　P. 3

Description			Materials	Labor	Subcontract & Equipment	Total
Vapor barrier	3347 ⌀	0.02	66			
Concrete	58 yds	$55	3190			
Finishing labor	3347 ⌀	0.16			535	
			4,740		3,541	8,287
Masonry						
Brick	11,000	$260/m	2,854			
Lintels	40lf		80			
Fireplace	prefab unit		690			
Hearth			85			
Mortar	80 sacks	4.50	360			
Sand	11yds	11.00	121			
Wall ties			68			
Stucco	460 ⌀	4.00			1840	
Masonry labor	11 m brick	210/m			2310	
			4,250		4,150	8,400
Framing						
PT plate	47 / 2x4x 14 P.T.	3.89	183			
Top plates & blkg	235 / 2x4x14	3.13	736			
Studs	800	2.70	2160			
Headers	4/2x12x10, 4/2x12x12, 8/2x12x14		271			
Corner Bracing	28 pc 1/2" CDX	9.75	273			

MFD IN U.S.A　　FRANK R. WALKER CO., PUBLISHERS, LISLE, ILLINOIS

Figure 3-2. Sample Estimate (Continued)

ESTIMATE WORK SHEET

JOB Lot 175 Bally Acres

P. 4

Description			Materials	Labor	Subcontract & Equipment	Total
Wall sheathing	60 pc 1/2" x 4 x 9	8.75	525			
Posts	4 / 6x6x10 PT	22	88			
Beams	3/2x12x12 , 2/2x12x14	9.36 & 10.92	97			
	glue lams		750			
Ceiling Jsts	55 /2x6x10	3.25	179			
	47 / 2x6x12	3.90	183			
	40 / 2x8 x14	6.52	261			
	13 / 2x12 x18	10.90	142			
	16/2x12 x 20	21.00	336			
Rafters	32 / 2x6x12	3.90	125			
	58 / 2x6x14	4.55	264			
	22 /2x6 x 18	6.28	138			
	8/2x6 x20	9.50	76			
	37/2x6 x 22	12.65	468			
Ridge	9/2x8x14	8.95	81			
Hip & valley	25/2x8x14	8.95	246			
Bracing	30/ 2x4x16	4.90	147			
Strong Backs	10 / 2x6x14	4.55	46			
Subfacia	25 /2x4 x14	4.55	114			
Roof sheathing	160 pc 1/2" CDX	9.75	1560			
Plywood clips	3 boxes	8	24			

MFD IN U.S.A. FRANK R. WALKER CO., PUBLISHERS, LISLE, ILLINOIS

Figure 3-2. Sample Estimate (Continued)

ESTIMATE WORK SHEET

JOB Lot 175 Bally Acres P. 5

Description			Materials	Labor	Subcontract & Equipment	Total
Felt	14 rolls	8	112			
Siding	1400 lf 1x6 lap	0.25	350			
Facia	360 lf 1x4	0.76	274			
Soffit	25 pc 3/8" ac	14.88	372			
Frieze	340 lf	0.34	116			
Vent	340 lf	0.47	160			
Ext locksets	3	30	90			
Windows			3680			
Trim boards	4/2x4x10, 6 pc 3/8 AC	2.50 & 14.88	99			
Ext doors			1400			
Shutters	8	25	200			
Framing labor	3347 ⌀	3.00			10,041	
			16,326		10,041	26,367
Roofing						
Roof edge	340 lf	0.20	68			
Valley flashing	140 lf	0.89	125			
Shingles	51 squares	37 & 25	1887		1275	
Flashing			100			
Nails	3 boxes	32.50	98			
Ridge	120 lf	0.67	80			
			2358		1275	3633

MFD IN U.S.A. FRANK R. WALKER CO., PUBLISHERS, LISLE, ILLINOIS

Figure 3-2. Sample Estimate (Continued)

ESTIMATE WORK SHEET

JOB Lot 175 Bally Acres P. 6

Description		Materials	Labor	Subcontract & Equipment	Total
Plumbing					
Basic package				6,125	
lavs		667			
Tub enclosures		650			
Shower door		320			
Sewer & water Tie in				420	
		1,637		6545	8,182
Electrical					
Basic package				4775	
Fixtures		2300			
Telephone prewire				75	
Cable prewire				80	
Alarm system				475	
		2300		5405	7,705
HVAC					
Basic package				6050	
				6050	6050
Insulation					
Walls	2472 Ø .23			621	
Attic/clg	2700 Ø 0.55			1360	
				1981	1981

MFD IN U.S.A. FRANK R. WALKER CO., PUBLISHERS, LISLE, ILLINOIS

Figure 3-2. Sample Estimate (Continued)

ESTIMATE WORK SHEET

JOB Lot 175 Baily Acres   P. 7

Description			Materials	Labor	Subcontract & Equipment	Total
Drywall						
Regular 1/2"	11,000 □		1,190			
M/R drywall	200 □		30			
Install					1,120	
Finish					1,792	
			1,220		2,912	4,132
Interior trim						
Base	900 lf	0.65	585			
Crown	714 lf	0.56	400			
Casing	425 lf	0.34	145			
Chair rail	50 lf	0.28	14			
Jambs	60 lf	0.74	44			
Window stool	70 lf	0.67	47			
Interior doors			1700			
Shelving	13/1x12x12	9.42	122			
Mirrors					313	
Locksets			170			
Door stops			15			
Bath Accessories			120			
Nails			150			
Labor					2472	

MFD IN USA   FRANK R. WALKER CO., PUBLISHERS, LISLE, ILLINOIS

Figure 3-2. Sample Estimate (Continued)

ESTIMATE WORK SHEET

JOB Lot 175 Bally Acres P. 8

Description			Materials	Labor	Subcontract & Equipment	Total
Interior trim			3,512		2,785	6,297
Paint & Wallcovering						
Paint	3347 ☐	2.00			6694	
Wallpaper			200		120	
			200		6814	7,014
Floor Covering						
Carpet	102 yds	20			2040	
Wood	540 ☐	5			2700	
Tile	1014 ☐	6			6084	
					10,824	10,824
Cabinets					11,655	
					11,655	11,655
Appliances						
Dishwasher			450			
Range & exhaust			775			
Oven			680			
Microwave			270			
Disposal			50			
			2225			2225
Exterior						
Final grade					450	

MFD IN U.S.A. FRANK R. WALKER CO., PUBLISHERS, LISLE, ILLINOIS

Figure 3-2. Sample Estimate (Continued)

ESTIMATE WORK SHEET

JOB Lot 175 Bally Acres

P. 5

Description		Materials	Labor	Subcontract & Equipment	Total
Top soil		200			
Trees		150			
Landscape				3000	
Sod		800			
Walks	400 ⌀ 5 yd conc	215		200	
Drive	2400 ⌀ 30 yd conc	1290		1200	
Sprinkler System				3000	
Cleaning				500	
		2655		8350	11,005
Subtotal		41,423	3,750	168674	213,847
Sales tax	41,423 x 0.07 7%	2900			2900
Labor burden	3,750 x 0.29 29%		1088		1088
		44,323	4838	168674	217,835
General overhead	217,835 x 0.062 6.2%				13,505
Total Cost	Direct Job cost plus general overhead				231,340
Profit	231,34 x 0.10 10%				23134
Sales Price					254,447

MFD IN USA

FRANK R. WALKER CO., PUBLISHERS, LISLE, ILLINOIS

Figure 3-2. Sample Estimate (Continued)

Figure 3-3. Sample Budget for Single-Family House

Category	Raw Cost	Final Budget (with Sales Tax and Labor Burden)
1. Land costs	$ 61,500	$61,500
2. Job overhead costs	27,250	28,338
3. Site work	1,340	1,340
4. Footings and slabs	8,287	8,619
5. Masonry	8,400	8,698
6. Framing	26,367	27,510
7. Roofing	3,633	3,798
8. Plumbing	8,182	8,296
9. Electrical	7,705	7,866
10. Heating, ventilation, and air-conditioning	6,050	6,050
11. Insulation	1,981	1,981
12. Drywall	4,132	4,217
13. Interior trim	6,297	6,543
14. Paint and wall-covering	7,014	7,028
15. Floor covering	10,824	10,824
16. Cabinets	11,655	11,655
17. Appliances	2,225	2,381
18. Exterior	11,005	11,191
	$213,847	$217,835

Note: The amounts in these columns would vary considerably in various parts of the country.

Post Ledgers

In keeping your cost records, you want to keep all the costs associated with a particular category of a certain job together so that they can be totaled and analyzed easily. One way to keep them together is to develop a ledger sheet or spreadsheet for each job. To use this method, you would list the categories across the page and each line down the page would be reserved for a cost item. Figure 3-4 shows a ledger for one job using this method.

You begin by listing the categories across the page along with a column for the date, the payee, and the check number. For convenience, some builders also list the amount and running subtotal. You can buy ledger sheets with 21 columns at office supply stores. However other blank ledger formats are also available if you need more or fewer columns. The 21-column ledger shown comes on one sheet, but it is presented here on three separate pages. The example given took two ledger sheets and is therefore shown on six pages. Note that the budget amount for each category is listed across the page for quick reference.

National® Brand 45-621 Eye-Ease®
Made in USA Lot #6 Bally Acres

Initials	Date
Prepared By	
Approved By	

				1	2	3	4	5
				Budget	217,835	61,500	28,338	1,340
	date	Payee	ch #	Amount	Subtotal	Land	Job Overhd	Site Work
1	1 3	Bally Acres Development	104	5700 00	5700 00	5700 00		
2	1 5	B.R. Plan Service	111	500 00	6200 00		500 00	
3	1 19	Gotham Title Co	–	53750 00	59950 00	51300 00	2450 00	
4	1 21	City of Gotham	119	470 00	60420 00		470 00	
5	1 27	Gotham Power & Light	120	25 00	60445 00		25 00	
6	1 27	Gotham Water Co	121	250 00	60695 00		250 00	
7	1 28	Jones Culvert Co	125	407 16	61102 16			407 16
8	2 19	Teague Conc. Pumping	130	486 00	61588 16			
9	2 19	David James Grading Co	146	1667 50	63255 66			1667 50
10	2 19	Roy's Carpentry Co.	147	1675 00	64930 66			
11	2 19	Slab Prep, Inc.	150	845 00	65775 66			
12	2 19	D&B	151	2040 00	67815 66			
13	2 20	Brent Brown Trucking	155	341 81	68157 47			
14	2 25	Termirid	161	502 00	68659 47			
15	2 25	Concrete Finishers Inc.	162	535 00	69194 47			
16	3 01	Ross Insurance Co.	170	197 00	69391 47		197 00	
17	3 05	Gotham Surveying Co.	176	200 00	69591 47		200 00	
	3 07	Roy's Carpentry Co.	180	3000 00	72591 47			
19	3 10	Home Lumber Co	186	1055 54	73647 01			
20	3 10	Bluff River Readi Mix	187	3310 96	76957 97			
21	3 10	Second National Bank	188	640 17	77598 14		640 17	
22	3 13	Ames Window Co.	193	3794 23	81392 37			
23	3 17	Roy's Carpentry Co.	199	3000 00	84392 37			
24	3 24	Roy's Carpentry Co.	206	3000 00	87392 37			
25	3 28	Roy's Carpentry Co.	210	1041 00	88433 37			
26	4 10	Home Lumber Co.	217	13201 06	101634 43			
27	4 10	Apex Brick Co	218	3570 17	105204 60			
28	4 10	Gotham Power & Light	219	21 41	105226 01		21 41	
29	4 10	Dynelectric	220	2425 00	107651 01			
30	4 10	Mubar Masonry	222	1320 00	108971 01			
31	4 12	Gulf States Roofing Co.	225	1200 00	110171 01			
32	4 12	Bob's Cleaning Service	226	150 00	110321 01			
33	4 12	Portatoilet, Inc	228	125 00	110446 01		125 00	
34	4 12	Brent Brown Trucking	229	134 17	110580 18			
35	4 17	Mubar Masonry	241	1050 00	111630 18			
36	4 18	Rockett Door Co	244	1244 61	112874 79			
	4 20	D&B Plumbing	247	2040 00	114914 79			
38	4 20	AAA Heating & Air	248	3025 00	117939 79			
39	4 20	NSG Security Co.	250	350 00	118289 79			
40	4 25	Drywall Hangers, Inc	256	1120 00	119409 79			
	4 25	Blue Ridge Insulators	257	610 05	120019 84			

Figure 3-4. Sample Job Cost Ledger

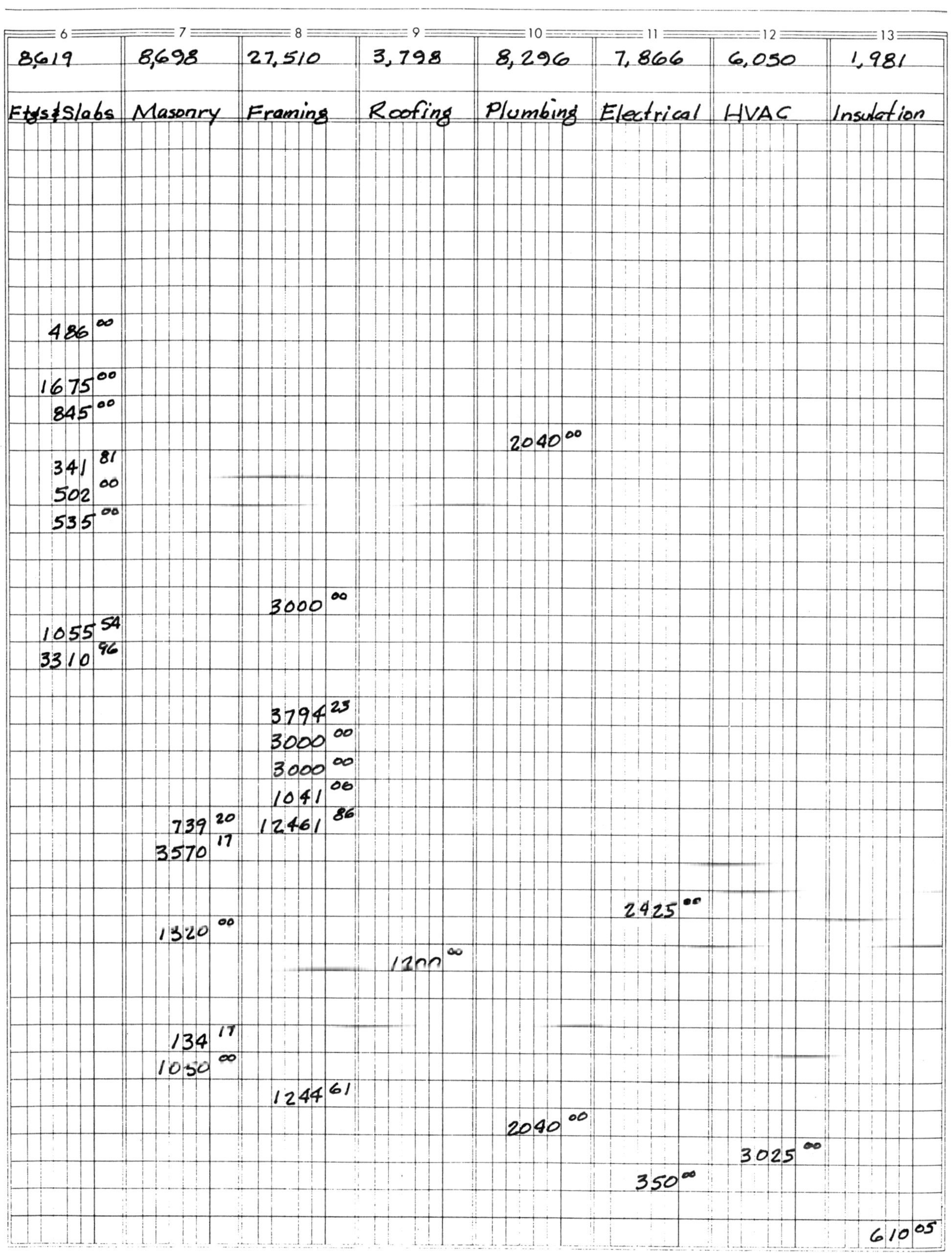

6	7	8	9	10	11	12	13
8,619	8,698	27,510	3,798	8,296	7,866	6,050	1,981
Ftgs&Slabs	Masonry	Framing	Roofing	Plumbing	Electrical	HVAC	Insulation
486.00							
1675.00							
845.00							
				2040.00			
341.81							
502.00							
535.00							
		3000.00					
1055.54							
3310.96							
		3794.23					
		3000.00					
		3000.00					
		1041.00					
	739.20	12461.86					
	3570.17						
					2425.00		
	1320.00						
			1200.00				
	134.17						
	1050.00						
		1244.61					
				2040.00			
						3025.00	
					350.00		
							610.05

Figure 3-4. Sample Job Cost Ledger (Continued)

	14	15	16	17	18	19	20	21
	4,217	6,543	7,028	10,824	11,655	2,381	11,191	
	Drywall	Interior Trim	Paint & Wallcovering	Floor Covering	Cabinets	Appliances	Exterior	Notes
1								
2								
3								Closing Statement
4								
5								
6								
7								
8								
9								
10								
11								
12								Rough in Slab
13								Porous fill
14								
15								
16								
17								
18								
19								
20								
21								Interest
22								
23								
24								
25								
26								
27								
28								
29								
30								
31								
32							150.00	
33								
34								Masonry Sand
35								
36								
37								Top out
38								
39								
40	1120.00							

Figure 3-4. Sample Job Cost Ledger (Continued)

National® Brand 45-621 Eye-Ease® Made in USA

Lot #6 Bally Acres

	Initials	Date
Prepared By		
Approved By		

date		Payee	Ch #	Amount (1)	Subtotal (2)	Land (3)	Job Overhead (4)	S/C Work (5)
				Budget	217,835	61,500	28,338	1,340
4	25	Drywall Finishers, Inc	258	1792.00	121811.84			
4	25	Second National Bank	259	927.12	122738.96		927.12	
5	04	Roy's Carpentry Co.	261	2472.00	125210.96			
5	09	Fixit Masonry Co.	267	1047.00	126257.96			
5	10	Apex Brick Co.	270	927.04	127185.00			
5	10	Home Lumber Co.	271	7506.76	134691.76			
5	10	Second National Bank	272	1017.61	135709.37		1017.61	
5	10	Wilson Stucco Co.	273	1836.00	137545.37			
5	18	Donald Ray's Paint Co.	285	6694.00	144239.37			
5	21	Hangit Wallpaper Co.	290	366.17	144605.54			
5	26	Larimat Cabinets	300	9226.00	153831.54			
5	27	Dalton's Carpet	301	2281.08	156112.62			
5	28	Bob's Cleaning Service	311	150.00	156262.62			
5	28	Underfoot Floor Co.	312	8704.00	164966.62			
5	28	Septic Systems, Inc.	313	4800.00	169766.62	4800.00		
5	28	David James Grading Co	314	325.00	170091.62			
5	30	Brent Brown Trucking	321	417.21	170508.83			
5	30	Concrete Finishers, Inc	322	1400.00	171908.83			
6	1	Ace Appliance Co.	328	2375.80	174284.63			
6	2	Blue Ridge Insulation	331	1344.00	175628.63			
6	2	NSG Security Co.	333	125.00	175753.63			
6	3	AAA Heating & Air	336	3025.00	178778.63			
6	3	Cultured Marble, Inc.	337	1400.00	180178.63			
6	3	Gotham Glass & Mirror	339	656.00	180834.63			
6	10	Gotham Power & Light	360	57.20	180891.83		57.20	
6	10	Ross Insurance Co.	361	75.25	180967.08		75.25	
6	10	Second National Bank	362	1321.17	182288.25		1321.17	
6	10	D & B Plumbing	363	2465.00	184753.25			
6	10	Dynelectric	364	2350.00	187103.25			
6	10	Cox Electric Supply	365	2493.17	189596.42			
6	10	Home Lumber Co	366	1304.17	190900.57			
6	12	Riley Landscape Service	371	3950.00	194850.57			
6	15	Payroll Account	375	4838.21	199688.80		4838.21	
6	15	Settlement Statement	–	15224.07	214912.87		15224.07	
6	15	Bluff River Concrete	377	1793.12	216705.99			
6	15	Bob's Cleaning Service	378	200.00	216905.99			
		Actual Expenditures			216905.99	61800.00	28339.21	2074.66
		Variance			929.01	<300.00>	<1.21>	<734.66>

Figure 3-4. Sample Job Cost Ledger (Continued)

6	7	8	9	10	11	12	13
8,619	8,698	27,510	3,798	8,296	7,866	6,050	1,981
Ftgs & Slabs	Masonry	Framing	Roofing	Plumbing	Electrical	HVAC	Insulation
	1047.00						
	927.04						
			3854.28				
	1836.00						
					125.00		1344.00
						3025.00	
				1400.00			
				342.40			
				2465.00			
					2350.00		
					2493.17		
8751.31	10623.58	27541.70	5054.28	8287.40	7743.17	6050.00	1954.05
<132.31>	<1925.58>	<31.70>	<1256.28>	8.60	122.83	0.00	26.95

Figure 3-4. Sample Job Cost Ledger (Continued)

Line	14	15	16	17	18	19	20	21
	4,217	6,543	7,028	10,824	11,655	2,381	11,191	
	Drywall	Interior Trim	Paint & Wallcover	Floor Covering	Cabinets	Appliances	Exterior	Notes
1	1792.00							
2								Interest
3		2472.00						
4								Teardown & rebuild wall
6		3652.48						Stolen Shingles
9			6694.00					
10			366.17					
11					9226.00			
12				2281.08				
13							150.00	
14				8704.00				
16							325.00	
17							417.12	Topsoil
18							1400.00	
19						2375.80		
24		313.60						Shower door & mirror
27								Interest
31	1[illegible]4.17							
32							3950.00	
33								Superintend. Salary & PT&I
35							1793.12	
36							200.00	
38	4216.17	6438.08	7060.17	10985.08	9226.00	2375.80	8385.33	
40	0.83	104.92	<32.17>	<161.08>	2429.00	5.20	2805.67	

Figure 3-4. Sample Job Cost Ledger (Continued)

As you approve each cost for payment or as you write the check for it, you would make the appropriate entry for the payment in the job ledger. For example, when you write the first check for Lot 6, Bally Acres (Figure 3-4), the date, the payee, the check number, and the amount is posted in the ledger. Because the first check (for $5,700) is paid to Bally Acres, Inc., for earnest money or a downpayment for the lot, the amount ($5,700) is posted under the land category. The builder would post the second check (to B. R. Plan Service for the plans) under job overhead. The third line represents the expenditures made at the lot loan closing. The total amount shown, $53,750, combines $51,300 for the balance owed on the lot and $2,450 for various closing fees charged by the attorney. These fees will appear on your closing statement. Because the closing costs are listed under the job overhead section, these charges are listed there. In all probability, you would not write a check for these expenditures, instead they would be charged to your construction loan.

At any point during the construction of the project, you can easily determine the expenditures to date for any category. By comparing the expenditures to date with those listed in the detailed estimate, you can tell whether or not you are within the budget for that category. Likewise, when the work associated with a certain category is finished, you can determine the final status of that category and determine the percent of variation.

From a management perspective, if a category runs over budget, the reason the category was over is more important than the amount. You can take a corrective action based on the reason. Focusing solely on the amount of the overrun does not provide enough direction toward improving your estimate and budget in the future. Be sure to cross check your change orders with your variance report. Some of what appears to be a variance may just be an unrecorded change order. (For a more detailed discussion on accounting for change order costs, see Chapter 6.)

For example, the final site-work bills were paid on February 19, and the expenditures totaled $2,074.66 or $734.66 more than the estimate). A quick check of the estimate shows that the estimator failed to include the cost of a culvert (which accounts for $407.16 of the overrun). The reason could have been that the estimator missed it altogether. Or possibly the originally planned location for the driveway was at the crest of a hill where the driveway did not need a culvert, and later the driveway location was moved. The grading and erosion control, done by David James, exceeded its budget by $327. This work was done by the hour, and hourly work often is difficult to estimate accurately. Controlling total cost is difficult when you are paying hourly wages because this arrangement carries no incentive for the workers and can vary for many reasons. And you still have to pay them even if adverse weather prevents them from doing the work productively. Again you must understand the reason for a variance and then take corrective action. If you do not understand a reason for a variance, you may not know what action to take to correct it.

If builders, remodelers, and developers keep a running check on exactly where a job's finances stand, the final tally should yield no surprises. At the conclusion of a job, you would total all the expenditures for all the categories and compare them to the budgeted amounts. These totals are shown on the line for actual expenditures. If you subtract the budgeted amount for each category from the actual expenditure, you will get the variance for each category (Figure 3-5).

The total job budget for the example shown is $217,835. This figure includes all direct costs to be incurred for the project. The final actual costs total $216,905.99. Thus the job came in $929.01 under the budget, off by less than half of one percent, which is close by anyone's reckoning.

However this percentage does not necessarily mean that the estimate was good or that the job went exactly as planned. The variances for each individual category ranged from $2,805.67 under the budget to $1,925.58

Figure 3-5. Method of Doing Variance Analyses

Category	Actual Cost	Budget	Variance	Comments
Land	$61,800.00	$61,500	–$300.00	Lower septic system bid found.
Job Overhead	28,339.21	28,338	–1.21	
Site work	2,074.66	1,340	–734.66	Driveway moved. Culvert needed for new location.
Footings and slabs	8,751.31	8,619	–132.31	Site was wet, and a conc. pump was used.
Masonry	10,623.58	8,698	–1,925.58	Mubar Masonry laid two walls badly and were paid. They refused to fix problems.
Framing	27,541.70	27,510	–31.70	
Roofing	5,054.28	3,798	–1,256.28	31 squares of shingles stolen.
Plumbing	8,287.40	8,296	8.60	
Electrical	7,743.17	7,866	122.83	
HVAC	6,050.00	6,050		
Insulation	1,954.05	1,981	26.95	
Drywall	4,216.17	4,217	0.83	
Interior trim	6,438.08	6,543	104.92	
Paint and wall-covering	7,060.17	7,028	–32.17	
Floor Covering	10,985.08	10,824	–161.08	
Cabinets	9,226.00	11,655	2,429.00	Found better bid.
Appliances	2,375.80	2,381	5.20	
Exterior	8,385.33	11,191	2,805.67	

Note: House sold before completion. Owners did not want sprinkler system.

over the budget. To get the full benefit of your cost control efforts, you should do a variance analysis for all the categories.

To do such an analysis, you simply compare the expenditures within a category to the estimate to see why they are different. Figure 3-5 shows one way to do the variance analyses. You may choose to figure the percent variance for each category. Likewise, many builders track and analyze the trends for each category which helps them to understand their businesses better.

Computerize Cost Control

At one time virtually all job ledgers were posted manually, and even today, many still are. However using a computer for job cost control can have many benefits. Basically the computer can do the work faster and with complete accuracy as long as the dollar amounts keyed in are correct. You can adapt most of the commercially available spreadsheet programs to do this work. Some programs on the market require only that you input the payee, the amount, and cost code. They will print the check and automatically post all the ledgers. Of course you have to set up the format and the codes to suit your own needs.

Many of the several programs for keeping and reconciling checking accounts can also track job costs. Figure 3-6 is a printout of the same information found on the manually printed ledger shown in Figure 3-4 but kept on a checking account program. Most of the checking account programs allow you to place your check in a category and print out all the checks written under that category. By selecting categories similar to those shown in Chapter 2, you can group all expenditures together for a given category, total them, and compare the total to a budgeted amount.

As computers become more common in the light construction industry, builders, remodelers, and developers look for software systems to help them run their companies better. Selecting software to suit a particular company requires time for a lot of research.

To make this job easier, the Software Review Program of the National Association of Home Builders researches and evaluates commercially available software for the light construction industry. If a particular software meets NAHB's rigorous performance standards that were developed by builder members, the Software Review Program grants it an NAHB Approved Software® Seal.

To achieve this distinction, the software must pass a series of controlled tests of its various functions, its performance of tasks, and its conformance to the Software Review Program's standards. NAHB evaluators use the software, survey other users, and conduct an evaluation of the vendor's training and support capabilities. Only after passing these tests, does the software earn the right to use the registered trademark seal of

Figure 3-6. Sample Spreadsheet Cost Control System

Itemized Category Report
1/1/94 Through 7/1/94

9/22/94 Page 1

Date	Acct	Num	Description	Memo	Category	Clr	Amount
	INCOME/EXPENSE						
	EXPENSES						
	aa Land						
1/3/94	Lot 6	104	Bally Acres D...		aa Land		-5,700.00
1/19/94	Lot 6		Gotham Title Co.		aa Land		-51,300.00
5/28/94	Lot 6	313	Septic System...		aa Land		-4,800.00
	Total aa Land						-61,800.00
	ab Job Overhead						
1/5/94	Lot 6	111	B.R. Plan Ser...		ab Job Overhead		-500.00
1/19/94	Lot 6		Gotham Title Co.		ab Job Overhead		-2,450.00
1/27/94	Lot 6	119	City of Gotham		ab Job Overhead		-470.00
1/27/94	Lot 6	120	Gotham Power ...		ab Job Overhead		-25.00
1/27/94	Lot 6	121	Gotham Water Co.		ab Job Overhead		-250.00
3/1/94	Lot 6	170	Ross Insuranc...		ab Job Overhead		-197.00
3/5/94	Lot 6	176	Gotham Survey...		ab Job Overhead		-200.00
3/10/94	Lot 6	188	Second Nation...	Interest	ab Job Overhead		-640.17
4/10/94	Lot 6	219	Gotham Power ...		ab Job Overhead		-21.41
4/12/94	Lot 6	228	Portatoilet, ...		ab Job Overhead		-125.00
4/25/94	Lot 6	259	Second Nation...	Interest	ab Job Overhead		-927.12
5/10/94	Lot 6	272	Second Nation...		ab Job Overhead		-1,017.61
6/10/94	Lot 6	360	Gotham Power ...		ab Job Overhead		-57.20
6/10/94	Lot 6	361	Ross Insuranc...		ab Job Overhead		-75.25
6/10/94	Lot 6	362	Second Nation...	Interest	ab Job Overhead		-1,321.17
6/15/94	Lot 6	350	Payroll Account	Super-sal./ P...	ab Job Overhead		-4,838.21
6/15/94	Lot 6		Settlement St...		ab Job Overhead		-15,224.07
	Total ab Job Overhead						-28,339.21
	b Site Work						
1/28/94	Lot 6	125	Jones Culvert...		b Site Work		-407.16
2/19/94	Lot 6	146	David James G...		b Site Work		-1,667.50
	Total b Site Work						-2,074.66
	c Footings&Slab						
2/19/94	Lot 6	130	Teague Concre...		c Footings&Slab		-486.00
2/19/94	Lot 6	147	Roy's Carpent...		c Footings&Slab		-1,675.00
2/19/94	Lot 6	150	Slab Prep, Inc.		c Footings&Slab		-845.00
2/20/94	Lot 6	155	Brent Browing...	Porous Fill	c Footings&Slab		-341.81
2/25/94	Lot 6	161	Termirid		c Footings&Slab		-502.00
2/25/94	Lot 6	162	Concrete Fini...		c Footings&Slab		-535.00
3/10/94	Lot 6	186	Home Lumber Co.		c Footings&Slab		-1,055.54
3/10/94	Lot 6	187	Bluff River R...		c Footings&Slab		-3,310.96
	Total c Footings&Slab						-8,751.31
	d Masonry						
4/10/94	Lot 6	217	Home Lumber Co.		d Masonry		-739.20
4/10/94	Lot 6	218	Apex Brick Co.		d Masonry		-3,570.17
4/10/94	Lot 6	222	Mubar Masonry		d Masonry		-1,320.00
4/12/94	Lot 6	229	Brent Browing...	Masonary sand	d Masonry		-134.17
4/17/94	Lot 6	241	Mubar Masonry		d Masonry		-1,050.00
5/9/94	Lot 6	267	Fixit Masonar...	Teardown & re...	d Masonry		-1,047.00
5/10/94	Lot 6	270	Apex Brick Co.		d Masonry		-927.04
5/10/94	Lot 6	273	Wilson Stucco...		d Masonry		-1,836.00
	Total d Masonry						-10,623.58

Figure 3-6. Sample Spreadsheet Cost Control System (Continued)

Itemized Category Report

9/22/94 — 1/1/94 Through 7/1/94 — Page 2

Date	Acct	Num	Description	Memo	Category	Clr	Amount
	e Framing						
3/7/94	Lot 6	180	Roy's Carpent...		e Framing		-3,000.00
3/13/94	Lot 6	193	Ames Window Co.		e Framing		-3,794.23
3/17/94	Lot 6	199	Roy's Carpent...		e Framing		-3,000.00
3/24/94	Lot 6	206	Roy's Carpent...		e Framing		-3,000.00
3/28/94	Lot 6	210	Roy's Carpent...		e Framing		-1,041.00
4/10/94	Lot 6	217	Home Lumber Co.		e Framing		-12,461.86
4/18/94	Lot 6	244	Rockett Door Co.		e Framing		-1,244.61
	Total e Framing						-27,541.70
	f Roofing						
4/12/94	Lot 6	225	Gulf States R...		f Roofing		-1,200.00
5/10/94	Lot 6	271	Home Lumber Co.		f Roofing		-3,854.28
	Total f Roofing						-5,054.28
	g Plumbing						
2/19/94	Lot 6	151	D & B Plumbing	Rough In Slab	g Plumbing		-2,040.00
4/20/94	Lot 6	247	D & B Plumbing	Top out	g Plumbing		-2,040.00
6/3/94	Lot 6	337	Cultured Marb...		g Plumbing		-1,400.00
6/3/94	Lot 6	339	Gotham Glass ...		g Plumbing		-342.40
6/10/94	Lot 6	363	D & B Plumbing		g Plumbing		-2,465.00
	Total g Plumbing						-8,287.40
	h Electrical						
4/10/94	Lot 6	220	Dynelectric		h Electrical		-2,425.00
4/20/94	Lot 6	250	NSG Security Co.		h Electrical		-350.00
6/2/94	Lot 6	333	NSG Security Co.		h Electrical		-125.00
6/10/94	Lot 6	364	Dynelectric		h Electrical		-2,350.00
6/10/94	Lot 6	365	Cox Electric ...		h Electrical		-2,493.17
	Total h Electrical						-7,743.17
	i HVAC						
4/20/94	Lot 6	248	AAA Heating &...		i HVAC		-3,025.00
6/3/94	Lot 6	336	AAA Heating &...		i HVAC		-3,025.00
	Total i HVAC						-6,050.00
	j Insulation						
4/25/94	Lot 6	257	Blue Ridge In...		j Insulation		-610.05
6/2/94	Lot 6	331	Blue Ridge In...		j Insulation		-1,344.00
	Total j Insulation						-1,954.05
	k Drywall						
4/25/94	Lot 6	256	Drywall Hange...		k Drywall		-1,120.00
4/25/94	Lot 6	258	Drywall Finis...		k Drywall		-1,792.00
6/10/94	Lot 6	366	Home Lumber Co.		k Drywall		-1,304.17
	Total k Drywall						-4,216.17

Figure 3-6. Sample Spreadsheet Cost Control System (Continued)

Itemized Category Report

1/1/94 Through 7/1/94

9/22/94 | Page 3

Date	Acct	Num	Description	Memo	Category	Clr	Amount
		l Interior Trim					
5/4/94	Lot 6	261	Roy's Carpent...		l Interior Trim		-2,472.00
5/10/94	Lot 6	271	Home Lumber Co.		l Interior Trim		-3,652.48
6/3/94	Lot 6	339	Gotham Glass ...		l Interior Trim		-313.60
		Total l Interior Trim					-6,438.08
		m Paint-Wallpap					
5/18/94	Lot 6	285	Donald Ray's ...		m Paint-Wallpap		-6,694.00
5/21/94	Lot 6	290	Hangit Wallpa...		m Paint-Wallpap		-366.17
		Total m Paint-Wallpap					-7,060.17
		n Floor Cover					
5/27/94	Lot 6	301	Dalton's Carp...		n Floor Cover		-2,281.08
5/28/94	Lot 6	312	Underfoot Flo...		n Floor Cover		-8,704.00
		Total n Floor Cover					-10,985.08
		o Cabinets					
5/26/94	Lot 6	300	Larimit Cabinets		o Cabinets		-9,226.00
		Total o Cabinets					-9,226.00
		p Appliances					
6/1/94	Lot 6	328	Ace Appliance...		p Appliances		-2,375.80
		Total p Appliances					-2,375.80
		q Exterior					
4/12/94	Lot 6	226	Bob's Cleanin...		q Exterior		-150.00
5/28/94	Lot 6	311	Bob's Cleanin...		q Exterior		-150.00
5/28/94	Lot 6	314	David James G...		q Exterior		-325.00
5/30/94	Lot 6	321	Brent Browing...	Topsoil	q Exterior		-417.21
5/30/94	Lot 6	322	Conc. Finishe...		q Exterior		-1,400.00
6/12/94	Lot 6	341	Riley Landsca...		q Exterior		-3,950.00
6/15/94	Lot 6	351	Bluff River R...		q Exterior		-1,793.12
6/15/94	Lot 6	352	Bob's Cleanin...		q Exterior		-200.00
		Total q Exterior					-8,385.33
	TOTAL EXPENSES						-216,905.99
	TOTAL INCOME/EXPENSE						-216,905.99

Figure 3-7. NAHB Approved Software® Seal

approval in Figure 3-7. The Software Review Program evaluates accounting, estimating, sales and marketing, and computer-aided design systems.

If you are curious about available software, you can get a copy of the *Software Directory for Builders and Remodelers*[5] and the *Software Review: Approved Product Summaries for Builders.*[6] The *Software Directory for Builders and Remodelers* lists over 300 software packages from 200 vendors with brief descriptions. The listings indicate the software that bears the NAHB Approved Software® seal. *Software Review: Approved Product Summaries for Builders* provides brief reviews and sample screens and menus for NAHB Approved Software® and includes the following items:

- a performance summary
- a checklist of features
- hardware requirements
- technical support available
- sample reports
- sample screens

You can obtain additional help from the following sources:

- educational programs at the NAHB Convention and Builders' Show, the NAHB Remodelers' Show, and other professional meetings including seminars and graduate level programs of the Home Builder's Institute
- Construction Financial Management Association in Princeton, New Jersey[7]
- the major software companies
- consultants

Make Operational Decisions

Analyzing various methods of doing some portion of the project or job allows you to compare the cost of paying your own employees by the hour for labor or equipment with the cost of having a subcontractor do it under contract. Many builders, remodelers, and developers find that keeping job costs properly helps them to make sound operational decisions based on past costs and projected costs. This information is the key to scheduling decisions as well as profit and cash flow projections.

Formalizing the analysis process by actually putting the numbers on paper as described above helps builders, remodelers, and developers to focus on identifying and preventing problems and on improving the quality of the estimate.

Develop Estimating Data Base

Perhaps the most important use of cost control techniques for builders, remodelers, and developers is the development and refinement of estimating data for planning future projects. For example, one of the most difficult categories to estimate properly is job overhead, especially those costs associated with loan closings. Different lenders have different items that end up on the list of loan closing costs. Likewise, different government-insured programs have varying categories of closing costs. Tracking these costs and comparing them to the estimate helps to ensure that all the many possible expenditures are considered in future estimates.

Enhance Cost Effectiveness

Builders, remodelers, and developers should continually evaluate various alternative systems, products, and technologies to increase their profits. Techniques for builders include the analysis of standard plans for speculative homes as well as the analysis of custom and semi-custom plans. Standard aspects of remodeler's jobs such as floor leveling should be analyzed frequently. Developers, too, should analyze alternatives such as various paving methods. All these efforts can be made easier if accurate historical cost records exist. For example, you may want to compare the cost benefits of using a computer-aided design system to the manual methods of producing plans. The use of value engineering techniques depends heavily on reliable, specific historic records for a favorable outcome.

Builders, remodelers, or developers who want to improve their existing cost control measures should take the steps listed in Figure 3-8.

Figure 3-8. Checklist to Track and Analyze Costs

- Divide typical work into categories.
- Establish a budget for each category from the estimate.
- Record each cost as it is incurred under the appropriate category.
- Analyze the costs periodically to identify and correct ongoing problems.
- Perform a variance analysis to determine the reason for any variances and your corrective action.
- Use information from present and past jobs to better plan new jobs and prevent future problems and mistakes.
- Use this information to decide whether to subcontract or do work in house.

Four

Track and Analyze General Overhead Costs

As mentioned previously, you can divide all the expenses or costs that a building, remodeling, or land development company incurs into two classifications: those costs that can easily be charged directly to a job and those that cannot. Those expenditures that can reasonably be allocated to a particular job are posted in that project's job cost ledger as described in Chapter 3. Those costs that cannot easily be charged to a job are called general overhead expenditures and are posted in the general overhead ledger. General overhead is sometimes referred to as office overhead or home office overhead.

The general overhead ledger tracks all indirect costs such as salaries of the builder's office employees and expenses associated with running the office. This general overhead ledger should not be confused with the general ledger. The general ledger is a financial accounting tool you use to keep track of expenditures and income for all of your various accounts. The general overhead ledger is a subsidiary ledger of the general ledger. You use the general ledger to prepare financial statements and profit-and-loss statements.

Set Up Cost Categories

In setting up a general overhead ledger, you must decide how to subdivide the expenses into main categories. *Accounting and Financial Management for Builders, Remodelers, and Developers* published by the National Association of Home Builders lists one way to subdivide the expenses in the General and Administrative section of the NAHB Chart of Accounts.[8] The 11 major categories are listed below and described in the following paragraphs:

- Salaries
- Payroll taxes and benefits
- Office expenses
- Telephone

- Computer expenses
- Vehicle, travel, and entertainment expenses
- Taxes
- Insurance
- Professional fees
- Depreciation expenses
- General and administrative, other

Salaries

For most small-volume builders, remodelers, and developers the salaries category will represent the largest single expenditure of the overhead pool. The businesslike way to handle your own income is to pay yourself a regular salary comparable to what you would have to pay someone else to do the same work. In this way, all the efforts that go into running the company are paid for as a company expense. If something should happen to you (illness or accident), you would have the budgeted funds to hire someone to fulfill your role. Some builders, remodelers, and developers do not pay themselves a fixed salary, instead they live off the so-called profits. Profit is what you have left over after all the expenses are paid, and your or anybody's time spent in running the business or making valuable contributions to it is just such an expense.

For example, suppose a building, remodeling, or land development company takes in just enough income to pay all the direct job costs (as defined in Chapter 3) and all the general overhead expenses. In this case the company could pay all its bills including the salary of the person or persons who run the company. However the company would make no profits to—

- make capital investments
- carry over to the next year
- provide working capital in lean times
- use to take advantage of unusual opportunities or to diversify
- pay bonuses

The owners of a building, remodeling, or land development company take the risks and should reap the rewards. Therefore, if a company makes a profit, the owners of the company can spend or use it as they please. Likewise, if a shortfall occurs and not enough money comes in to cover all the job costs and general overhead expenses, the first category to be cut is the money paid to the company's owners.

The salaries category includes the salaries paid to each employee whose efforts are not reasonably chargeable to a project or job. These personnel include those in the following list. In your firm one person may cover more than one of these positions:

- estimators
- purchasing agents or managers
- drafters, designers, and architects
- midlevel administrative personnel
- secretaries and clerks
- bookkeepers and in-house accountants
- salespersons on salary as opposed to commission
- nonworking superintendents

Payroll Taxes and Benefits

This category covers the company's payroll taxes and insurance—sometimes called labor burden—associated with the salaries of the people covered in the salaries category described above. You would charge to each individual job the labor burden expenses of all craftspeople, superintendents, and others who worked on that job. Items in this category include the company's share of the costs for all office personnel for—

- Social Security and Medicare
- federal and state unemployment insurance
- workers' compensation insurance premiums
- health and accident insurance premiums
- disability and life insurance premiums
- contributions to retirement or profit-sharing plans
- holiday and vacation pay
- any other fringe benefits relating to salaries and wages for administrative personnel

Office Expense

Except as outlined elsewhere in the general overhead expenses, this account would cover expenses associated with running the administrative and sales office as opposed to a field or site office. If you are renting or leasing office space, you would post those charges here. You would also include in this category any rental or lease charges on office equipment as well as repair and maintenance costs, costs associated with uncapitalized building repairs and remodeling expenses, utility costs, and janitorial services for this office.

You also would charge office supplies, telephone expenses, beepers, an answering service, and cellular phones to this category. Some builders, remodelers, and developers charge all long-distance expenses to general overhead while others separate out those charges that benefit particular jobs. This report can be tedious unless you have a phone system that requires an access code for long-distance calls. If each individual job is given its own code, you will find the allocation of long-distance charges easier.

Computer Expenses

All costs associated with your office computers that do not have to be capitalized fit into this category. This grouping includes computer hardware and software leases and maintenance contracts, paper, toner, ribbons, and miscellaneous supplies.

Vehicle and Travel

This category includes lease or rental payments on vehicles used by administrative personnel, repair and maintenance costs, fuel costs, and mileage reimbursements paid to administrative personnel for the use of their private vehicles. Other expenses that belong in this category include—

- taxes, licenses, fees, and insurance costs for the company's vehicles
- reasonable travel expenses for administrative personnel on company-related business that you cannot charge to a job
- education expenses such as in-house programs and the cost of travel to an outside event, lodging, registration fees, materials, and literature.

Entertainment Expenses

Because of changes in the Internal Revenue Service treatment of entertainment expenses, you might want to keep these charges separate. (They used to be included in vehicle and travel expense.) This category would include any entertainment expenses that are necessary to carry out a business activity.

Taxes

Under this category are real estate taxes such as tax on property used for the firm's offices and real estate taxes that you cannot charge elsewhere. This category would cover any assessments or taxes on personal property owned by the firm as well as any licenses, registrations, municipal fees, and operating permits. It also would include use taxes for out-of-state purchases of goods or services taxable in your home state.

Insurance

The cost of the insurance premiums that are part of this category include—

- life insurance (if not included in labor burden, whether or not it was deductible for federal income tax purposes)
- disability insurance (if not included in labor burden, whether or not it was deductible for federal income tax purposes)
- fire and extended coverage on buildings and contents not associated with a particular job
- property damage and liability insurance premiums on the firm's vehicles not charged elsewhere

- general liability insurance, including product liability insurance
- business continuation insurance
- umbrella policies
- other insurance that cannot be charged to another specific category or a project

Professional Services

This group of categories covers the cost of outside accounting services, including audits, tax assistance, and preparation of financial statements and profit-and-loss statements. It would also include subcategories for (a) fees for legal and other professional services that benefit the company as a whole rather than a particular project, (b) expenses associated with hiring administrative staff members and employees whose services are to be used for an indeterminate number of projects (such as salespeople and bookkeeping, clerical, or secretarial help), and training and education expenses (such as registration fees for seminars and conferences) plus other associated expenses.

Depreciation Expenses

Purchases you must charge as an expense over several years (rather than during a single fiscal year) represent a special classification of expense known as depreciation. If your company owns any buildings (such as its administrative office), you would charge their depreciation expenses to this category as well as depreciation on company-owned vehicles, furniture, fixtures, computer hardware and software, office machines, and other equipment. If you have expended money to improve office buildings leased from others, you may charge the amortization (depreciation) of these improvements to this category. It would also include any organizational costs you may need to amortize, including legal fees and corporate charter fees.

Many builders, remodelers, and developers charge depreciation of construction equipment to the general overhead pool. An investment in construction equipment that is spread over several years establishes a certain capacity or volume of construction that a builder, remodeler, or developer can do. Builders, remodelers, and developers try to maintain a level of activity that will support their investments in equipment. However depreciation costs continue to accrue regardless of the level of construction activity. Therefore these costs are independent of how much work is done and are often part of the general overhead.

General and Administrative Expenses, Other

In this category you will include such expenses as bad debts, charitable donations, dues and subscriptions to trade journals and other business publications (such as Home Builder Press books), and other charges not otherwise classified.

Post the General Overhead Ledger

Figure 4-1 shows a manually posted general overhead ledger. This ledger is kept for the fiscal year. In this example the fiscal year runs from January 1 to December 31.

Analyze General Overhead Expense

Listed across the top are the various categories with the budget for each category below. As general overhead expenses are incurred, you would post them as shown in the example. You should constantly compare the expenditures to the budget to determine whether your general overhead is ahead of or behind the expected, planned-for amounts. The key idea is that excessive general overhead eats into profits, and the best way to control overhead expenditures is to constantly monitor costs and compare them to budget and/or historical data.

Of course the general overhead ledger may be kept on a computer using a spreadsheet program or a checking account program. A sample is not shown here, but it would look similar to the computerized job cost ledger shown in Figure 3-4.

On the last page of the estimate shown in Figure 3-1, the cost is increased by 6.2 percent to cover the general overhead expenses that must be absorbed by this project. Many builders, remodelers, and developers will find that their general overhead may be considerably higher than this number while others may find their's even lower. The percentage depends on your organization and method of allocation. The best strategy is to allocate as much as possible to the individual jobs.

This overhead percentage is derived by dividing the total general overhead for the year by the total volume of work for that year. In this instance the total volume was estimated to be $1,370,000 and the total general overhead expenses were budgeted at $85,170. Each job has to bear its proportionate amount of the general overhead expenses. Therefore, if a job represents a fourth of the volume, that job must absorb a fourth of the overhead expenses.

For example, assume that a builder, developer, or remodeler enters into a contract to construct a single project whose direct cost is $210,000. All of the materials, labor, equipment, subcontracts, and job overhead total $210,000. Further assume that the project will take exactly 1 year and represents all the work that the builder will be doing during the year. If the builder estimates that general overhead for the year will be $25,000 and adds this amount to the direct job costs, that amount brings the total of all costs to the builder up to $235,000. If the builder determines that a 10-percent profit would be reasonable, that percentage would add another $21,000 to bring the contract price to $256,000. Of course any overrun or underrun in the direct costs of the job will decrease or increase the profits.

National® Brand 45-621 Eye-Ease®
Made in USA

General Overhead
1995 - 1996

	Initials	Date
Prepared By		
Approved By		

date		Payee	Ck #	1 Amount	2 Subtotal	3 Salaries	4 PT & I	5 Office Expenses
				Budget	85,170	49,320	10,350	13,000
1	5	Dewey, Cheatham & Howe	105	300 00	300 00			
1	5	O'Conner Realty	106	650 00	950 00			650 00
1	5	State Contractors Licensing Bd	107	100 00	1050 00			
1	10	Creditcard Corp.	112	126 12	1176 12			
1	10	Gotham Power & Light	113	176 24	1352 36			176 24
1	10	Central Bell	114	117 47	1469 83			117 47
1	10	Software, Inc.	115	78 21	1548 04			
1	10	Office Supply, Inc.	116	78 91	1626 95			78 91
1	27	Builder's World Magazine	122	24 00	1650 95			
1	28	Ross Insurance Co.	124	427 50	2078 45			
1	31	Tom Builder	127	2850 00	4928 45	2850 00		
1	31	Leslie Booker	128	1260 00	6188 45	1260 00		
1	31	Office of Employment Security	129	202 60	6391 05		202 60	
1	31	Second National Bank	131	319 76	6710 87		319 76	
1	31	Ross Insurance Co.	132	340 20	7051 01		340 20	
1	31	Lease-a-lot	133	365 00	7416 01			
		O'Conner Realty	140	650 00	8066 01			650 00
		... Light	142	131 20	8197 21			131 20
				42 17	8239 38			
					8266 ...			
								97 16
12	6	...						
12	8	O'Conner Rea...						
12	10	Gotham Power & Light	840					
12	10	Creditcard Corp	841	106 14	...			
12	10	Central Bell	842	97 16	76751 21			
12	18	LSU Dept. of Const.	856	500 00	76251 21			
12	20	City of Gotham	860	45 00	77296 21			
12	22	American Screenprinters	865	125 00	77421 21			
12	28	Bank Charges	-	6 28	77427 49			
12	31	Lease-a-lot	870	365 00	77792 49			
12	31	Tom Builder	871	2850 00	80642 49	2850 00		
12	31	Leslie Booker	872	1260 00	81902 49	1260 00		
12	31	Equipment Depreciation	-	2612 41	84514 90			
12	31	Counts & Talley CPA	873	421 17	84936 07			
		Actual Expenditures			84936 07	49320 00	10350 72	12816 42
		VARIANCE			233 93	0	<72>	186 58

Figure 4-1. Sample Manually Posted General Overhead Ledger

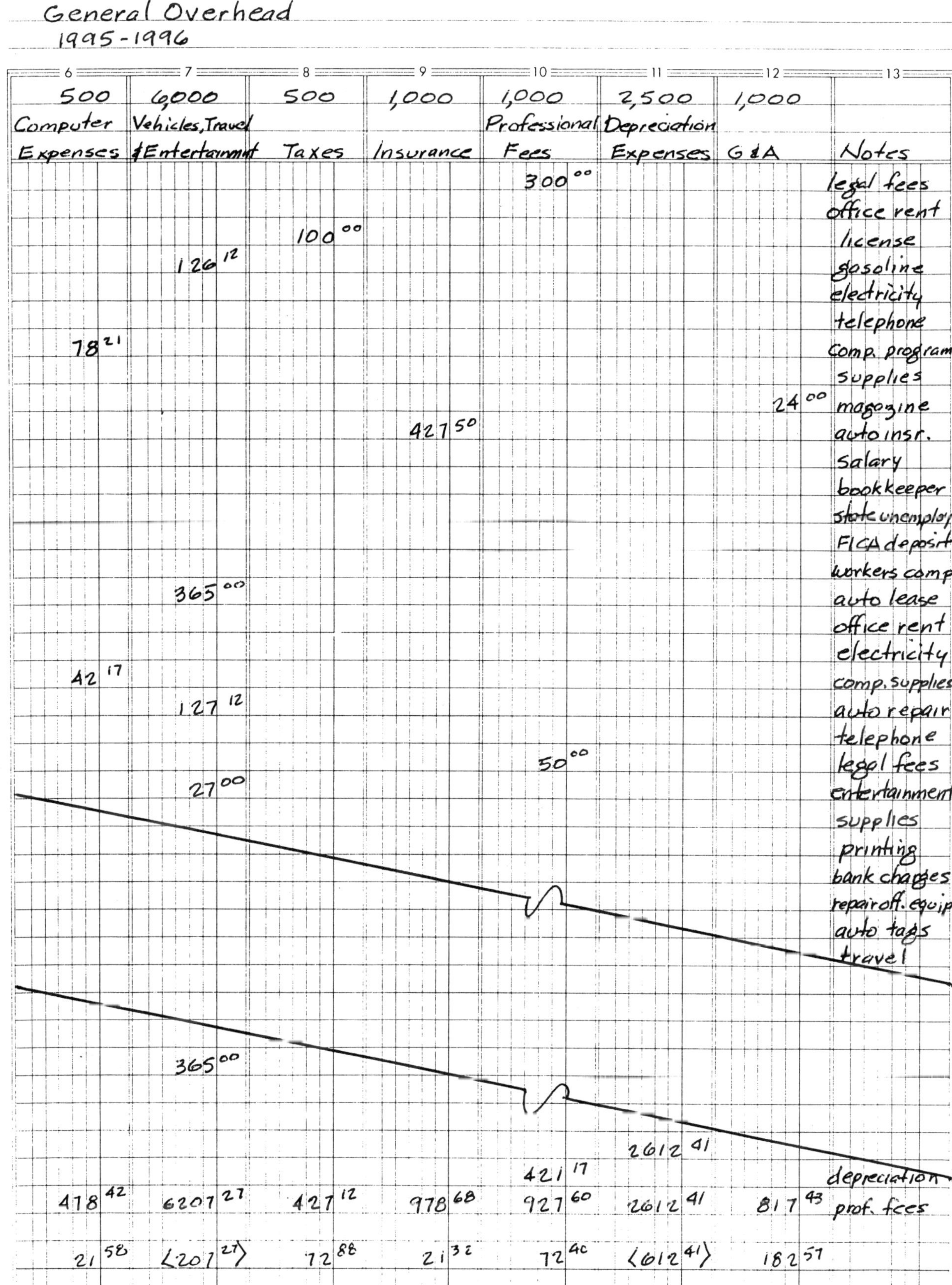

General Overhead
1995-1996

6	7	8	9	10	11	12	13
500	6,000	500	1,000	1,000	2,500	1,000	
Computer Expenses	Vehicles, Travel & Entertainment	Taxes	Insurance	Professional Fees	Depreciation Expenses	G & A	Notes
				300.00			legal fees
							office rent
		100.00					license
	126.12						gasoline
							electricity
							telephone
78.21							comp. program
							supplies
						24.00	magazine
			427.50				auto insr.
							salary
							bookkeeper
							state unemploy.
							FICA deposit
							workers comp
	365.00						auto lease
							office rent
							electricity
42.17							comp. supplies
	127.12						auto repair
							telephone
				50.00			legal fees
	27.00						entertainment
							supplies
							printing
							bank charges
							repair off. equip
							auto tags
							travel
	365.00						
					2612.41		
				421.17			depreciation
418.42	6207.27	427.12	978.68	927.60	2612.41	817.43	prof. fees
21.58	<207.27>	72.88	21.32	72.40	<612.41>	182.57	

Figure 4-1. Sample Manually Posted General Overhead Ledger (Continued)

Likewise, any overrun or underrun in the general overhead will decrease or increase profits.

The term *mark-up* describes how much a builder, remodeler, or developer increases or marks up the price of a job above its direct cost. In the example presented above, the direct cost was $210,000. The sales or contract price was $256,000. Therefore the mark-up was $46,000. Part of the mark-up goes to cover general overhead expenses and part of it is the profit. In this case the mark-up (general overhead and profit) was 21.9 percent ($46,000 ÷ $210,000). The general overhead was 11.9 percent ($25,000 ÷ $210,000) and the profit was 10 percent ($21,000 ÷ $210,000).

To plot a secure financial course, a builder, remodeler, or developer must be able to predict the general overhead costs accurately and distribute them to the projects.

Tracking and analyzing these expenses using the methods described in this book will help builders, remodelers, or developers to more accurately estimate their total costs, and it will help to increase their awareness of these expenditures and, therefore, help to keep them under control.

Of course, you want to do as much sales volume as possible using the fixed general overhead expenses you have. A company that is too top heavy—general overhead expenses too high—has trouble surviving.

In summary the proper way to treat general overhead is itemized in Figure 4-2.

Figure 4-2. Checklist for Tracking and Analyzing General Overhead

- Establish categories into which you place all general overhead expenses
- Charge only those items to general overhead that you cannot reasonably charge to a job
- Constantly monitor general overhead expenses to see if you can find or develop some more cost-effective method to accomplish the same purpose
- Determine the percent of your expenditures that general overhead represents
- Use this percent to mark up your estimates to accurately disperse general overhead costs to the jobs
- Use this percent as a guide to price changes in existing work
- Expand or contract general overhead expenses to meet the demand for work

Five

Operate Your Purchasing System

In order to have an efficient cost control system, a builder, remodeler, or developer must control the cost of materials, labor, equipment, and subcontracts for the various jobs.

Materials

In controlling material costs, the builder, remodeler, or developer must constantly monitor both the quantities of materials and their unit prices. Cost control begins during the design phase of a job as you produce or choose economical, cost-effective designs. For example, a home with four bathrooms and two fireplaces will usually cost more per square foot if it is 2,000 square feet rather than 3,000 square feet because the extra cost of these high-priced items is spread over a larger area.

During the construction phase, the builder, remodeler, or developer can control material costs with the judicious use of an efficient purchasing system. In really small-volume companies, the purchasing function is usually handled by the owner of the firm, except for an occasional small item bought to fill an immediate need. Large-volume organizations usually need a more sophisticated purchasing system than a small-volume company. However even the smallest firms can benefit from the use of somewhat more formalized purchasing procedures.

No purchases should be made without going through the proper channels. If purchases are unauthorized, the same materials may be ordered twice, which increases the likelihood of losses from theft, spoilage, and wasted materials.

Even with a purchase order system, shortages of materials may develop for several reasons. Theft, spoilage, back-orders, an inaccurate estimate, and incorrect use of materials are only a few. Because much (if not most) of the labor is subcontracted, the shortage is often discovered by a subcontractor. Many subcontractors will tell you exactly how many of an item to send out to finish the work. Even if the subcontractor is a good estimator, you should not order based on his or her calculations. You should still do your own estimate as a check against possible mistakes. Thus you avoid (a) having to reorder to fill a later shortage or (b) deal with a surplus that creates waste and increased costs.

Typically the person who initiates the purchase request is the individual in charge of actually building the job or in a medium- or large-volume company, the superintendent. Based on the estimate for the job, he or she prepares the takeoff or list of materials that are needed for the job. Small-volume builders and remodelers commonly place orders through their lumber salesperson by telephone or fax or over the counter at the lumber company. As long as the company remains small enough, this procedure usually is adequate because the person ordering can remember quotes and other details. However, as the operation becomes larger, these informal methods may not prove satisfactory. One Midwest builder has such a good relationship with his lumber salesperson that the salesperson calls each of his jobs every day to see what is needed and plan the next delivery.

Purchasing Objectives

No matter what the size of the organization, the person in charge of purchasing needs to be aware of the major functions or objectives to be achieved by the purchasing process that are listed below:

- Obtain competitive price.
- Maintain quality control.
- Obtain prompt and complete delivery.
- Verify quantity and condition of items delivered.
- Compare invoiced cost to the estimate.
- Pay promptly.

A checklist for operating your purchasing system appears in Figure 5-1.

Figure 5-1. Checklist for Operating Your Purchasing System

- Determine a system for obtaining the best prices.
- Establish a workable quality control system.
- Develop procedures to ensure prompt and complete deliveries.
- Set up a system to check quantity and quality for materials.
- Check invoices against quotes or purchase orders.
- Establish a list of qualified, reliable subcontractors for each trade.
- Set up policies and procedures for warranty work, back charges, and call backs.
- Develop a comprehensive set of policies to cover all aspects of direct labor employees.
- Make sure that your equipment purchases take all important factors into consideration.

Obtain Best Price

The first objective in setting up your purchasing function is price. Many builders are able to increase their profits substantially by searching for more competitive prices. One way of getting better prices is to prepare a list of all the materials that you typically use on your projects. Send this list to all the material suppliers in the area with whom you might deal and ask them to quote the price of each item. For you to compare prices accurately, ask each supplier to specify the time period for which the prices are good.

Some builders buy materials from more than one lumber company while other builders find it more profitable to buy all of their major materials for a particular project from the same supplier. Buying most of your materials from one company, but some from one or two others can give you some protection if something happens to the first firm or if it runs out of something unexpectedly. You have an established relationship to fall back on. Always ask for and deal with the same person when you place orders so that person gets to know you and your company. Having an open account at one or two large building supply stores that serve homeowners directly will often take care of emergencies.

While price is an important factor, the savings you might achieve from lower prices can be offset by losses suffered because of delays and substandard quality that can often result in a higher final cost. A builder or remodeler who spends an inordinate amount of time correcting or compensating for the mistakes of others cannot produce as much as another person who does not have such problems. The smoother your operation, the more work you can produce.

Maintain Quality Control

The second objective of purchasing is to ensure that the proper quality control is achieved. Decisions about quality are important to builders for many reasons. If you are building a house or doing a remodeling job on a contract basis, the specifications usually spell out the kinds and quality of material needed. The owner may not always end up being happy with the selections even though they are set forth in the contract.

Most home buyers and homeowners are not thoroughly familiar with all the products that make up a house or remodeling project. Therefore the builder or remodeler should explain and demonstrate the quality of the materials for the owner as thoroughly as possible prior to purchasing them. A good way to demonstrate the quality of the materials is to show the owner what the materials look like installed in another house. Builders find that different owners have different tastes. For example, some owners are perfectly happy with aluminum windows while others are satisfied only if they have wood windows. Home buyers and homeowners have so many choices to make that many times they do not realize they are going to be dissatisfied with a particular product until they see it or see it installed.

In choosing the quality of materials for affordable speculative houses, you may want to choose less-expensive materials but not to the point that it affects your sales. For example, in a high-end market, if you put hollow-core, flush-panel, interior doors in a $250,000-dollar speculative home, you might not sell the house as quickly as if you had used a more expensive style. However, you can also possibly select quality that is too good for a particular price range. Potential buyers may not recognize the extra value of the materials nor be willing to pay what that premium is worth.

Obtain Prompt and Complete Delivery

When you are making purchasing decisions, you need to consider your supplier's ability to deliver the required materials on time. Delays in the delivery of materials can disrupt the job and cause unforeseen expenses. Further, delivery of materials before they are needed often leads to theft and damage or deterioration of the materials.

Verify Quantity and Condition of Delivery

Ideally every delivery made would be on time with the quantities correct, and all items would be in excellent condition. However deliveries are sometimes less than ideal. Therefore you need to consider the willingness of a supplier to be reasonable when errors occur. If your material supplier delivers inferior materials, charges prices higher than you were quoted, or lists quantities in excess of those delivered, you expect the supplier to be reasonable and correct any errors without a lot of arguments.

Use Purchase Orders

The purchase order is a contract that specifies the quantity and quality of materials to be supplied, the unit price, the total price, and other terms of the sale. The purchasing function actually starts when you contact suppliers and receive a quote for the needed materials. When you need the materials, you prepare a purchase order for a particular delivery using the prices quoted by the supplier. The person in your company who pays the bills gets one copy. When the invoice arrives, that person should compare the invoice with the purchase order to determine the accuracy of the invoice. A second copy of the purchase order goes to the job where it becomes a checklist for incoming materials.

Using purchase orders prevents unauthorized purchases. All purchase orders should be numbered sequentially and the numbers controlled through the use of a purchase order register. The purchase order register is simply a record of purchase orders for a job in numerical sequence. The register should list each purchase order number, the vendor's name, and the total amount committed.

To properly control spending, you should limit the authority to issue purchase orders to a comparatively small number of individuals. Purchase orders should be carefully prepared with special attention to the details. All the conditions of the sale including material specifications, quantity, unit price, delivery time, destination, taxes, discounts, and other specific terms should be written on the purchase order. Figure 5-2 shows a typical purchase order.

Expedite Deliveries

A builder or remodeler cannot automatically assume that all the delivery dates specified in the purchase orders will be met. A simple way of expe-

Figure 5-2. Sample Purchase Order

Purchase Order

CENTURY CONSTRUCTION COMPANY
Post Office Box 81743
Atlanta, Georgia

Date 9/10/95 Date Required 9/15/95

Ship Via Your Truck

To Home Lumber Company

Ship to Lot Number 6

" Bally Acres S/D

Quantity	✓	Stock Number/Description	Price	Per	
900 lf		Colonial Base, C Fir	.65	foot	$ 585.00
714 lf		3 1/2" Crown, C Fir	.56	foot	399.84
425 lf		2 1/4 Casing, C Fir	.34	foot	144.50

Subtotal $1,129.34

TERMS: 2% 10 Days
Net 30 Days

TAX 79.05

TOTAL $1,208.39

Important -- our number must appear on invoices, packages and correspondence.
Acknowledge if unable to deliver by date required.

diting material deliveries is to call the supplier to confirm that all the materials are available and that the supplier will meet the delivery schedules.

Expediting standard off-the-shelf materials usually presents few problems. However the delivery of items specially fabricated for the job (such as custom cabinets, special roof trusses, or special entry signs) may require one or more check-back calls to ensure their timely delivery.

Monitor Back Orders

One purchasing problem builders, remodelers, and developers sometimes must solve is the situation in which the supplier is unable to deliver all of the items requested on the purchase order. The supplier delivers part of the order and places the missing items on backorder. Of course, if the builder or remodeler has carefully determined what items are available and put them on hold at the supplier, this situation happens less frequently and is potentially less costly.

When certain items are backordered, you might want to purchase them elsewhere to avoid production delays. If you do that, be sure to cancel the backordered items in writing. Unfortunately such items arrive after the new items have been installed and cause problems. Usually you would do better to cancel the old purchase order and write a completely new one that specifies only the items the supplier has, rather than hold the old purchase order open and wait for the materials to arrive. This procedure helps to keep your records straight and avoids double delivery.

Subcontractors

The current trend in residential construction is for small- to moderate-size companies to use subcontractors to a greater degree. For many years builders and remodelers did the vast majority of their work with their own hourly employees. Their reputations depended directly on the competence of the craftspeople they employed. Today some builders and remodelers as well as many developers, commonly subcontract 100 percent of their work. This practice allows them more flexibility, reduces their risk, and increases their profits (Figure 5-3).

Using subcontractors has increased in popularity to the point that a builder's, remodeler's, or developer's supply of reputable subcontractors can mean the difference between success and failure. In residential work, more so than in other forms of construction, the relationship between a builder, remodeler, or developer and a subcontractor tends to be more ongoing than transient. Builders, remodelers, and developers often use the same subcontractors again and again. For example, many of them who are pleased with the performance of a particular subcontractor may try to retain him or her on the team for the long term. Furthermore, if a builder, remodeler, or developer and a subcontractor have a sound work-

Figure 5-3. Reasons for Using Subcontractors

- Using subcontracted labor makes the builder's operations more flexible. The builder can cut back during times when business is slow and expand during times when demand is high.
- Subcontractors specialize in certain trades (such as framing; heating, ventilation, and air-conditioning; electrical; plumbing; and drywall). Their workers usually are more skilled than the those who do the work periodically.
- Often the more highly skilled subcontracted crew can do the tasks involved more economically than the builder's or remodeler's own people who are less specialized.
- Many tasks require special, expensive equipment that the subcontractor can buy and use everyday, especially heavy, earthmoving equipment, scaffolding, and specialty tools. Such equipment purchases are usually not cost effective for most small-volume builders, remodelers, and developers.
- Many subcontractors are small organizations in which the craftspeople are either related to one another or have worked together for a long time. They are accustomed to working together efficiently on the job.
- Subcontractors typically work by the job or on some basis other than by the hour. Because working as efficiently as possible is in their best interest, a builder, remodeler, or developer can save money by not having to provide continuous supervision to ensure good productivity. The subcontractor also bears the cost of callbacks.
- Subcontractors may be able to purchase certain materials at better prices because of the larger quantities they buy.

ing relationship, they become familiar with each others' policies and practices and the work can progress more smoothly. Such a relationship usually also improves productivity and quality.

Even under the best of conditions, in which a builder, remodeler, or developer has a team of steady, reliable subcontractors, he or she occasionally may need to find new ones. Perhaps the best way to find new ones is to talk to members of your local builders association or your Remodelors® Council, suppliers, lenders, and others in your professional network.

One of the easiest ways to find new subcontractors is to travel the area where a concentration of construction is in progress and find the subcontractors at work. This practice allows you to observe the quality of their work. Besides the quality of work, of course, you must consider the reliability of subcontractors and their financial stability. Other builders, remodelers, and developers; your suppliers; and others in the industry can often recommend good subcontractors.

A subcontractor should make a reasonable profit. If the subcontractor's profit is too great, the excess comes out of the builder's or remodeler's profit. If the subcontractor's profit is inadequate, quality may suffer or the subcontractor could even go out of business, and possibly delay the completion of the current job and complicate planning for the next one. Warranty work might be delayed, and it also could be more expensive.

Analyze Subcontractor Bids

When evaluating a subcontractor's quote you should consider the exact scope of work proposed. When you are using a number of subcontractors, you have to be careful that you (a) leave no gaps in assigning responsibility for work and (b) create no overlaps in responsibility. For example, if you need a disconnect at an electric heat pump, the electrician may supply it or the heating, ventilation, and air-conditioning subcontractor may do so. A gap would result if neither supplied it, and an overlap would result if both planned to provide it.

Monitor Standard Subcontractor Prices

When analyzing subcontractors' prices, you should consider that what appears to be the lowest bid might not be the best bargain. Problems with a subcontractor whose work is unsatisfactory can often cost more than the difference saved. If you use trustworthy, reliable, high-quality subcontractors, often you can produce considerably more and better work than a builder, remodeler, or developer who does not use them.

Method of Pricing—Many subcontractors will do their work on a unit-price basis. For example, framing subcontractors, concrete slab finishers, and painters often base their bids on square footage of floor area. Masons lay brick by the unit, roofers install shingles by the square, and drywall hangers hang by the square foot of board. You can easily compare the prices of these subcontractors with those of others who bid the same way. Other subcontractors may quote only lump-sum prices (with a base price and a listing of add-on prices for extras). Plumbers, electricians, and heating, ventilation, and air-conditioning subcontractors usually bid this way.

Competitive Prices—You have two ways to check or monitor your standard or sole-source subcontractors' prices to make sure that they are still competitive. You can estimate the job yourself, or you can get another subcontractor to bid the work.

For instance, imagine that in your area all the subcontractor prices for landscaping are inordinately high. You can easily determine the quantities and prices of material and labor used on a typical job and thereby the profit. Such a situation may provide an opportunity for you to go into another business. If your own work load is as much as you want to handle, you may be able to find a competent person to put into business with whom you can share the profits.

You should incorporate a separate business entity for these purposes. One builder who occasionally bid small commercial jobs in his area did several projects with single-ply, flat roofs. The roofers were all rather expensive. Therefore at the urging of a roofing manufacturer's representative, this builder went into the single-ply roofing business on his own projects. He made as much profit on the roof as he did on the rest of the project. Unfortunately, the glue deteriorated and the roofs leaked. The manufacturer's company-underwritten roofing bonds became worthless. The builder had to replace the roofs and repair the resulting water damage, and the costs bankrupted him. This builder's experience provides several lessons, including (a) do not go into businesses you do not fully understand and (b) protect your main business from being pulled down by a side venture.

Workers' Compensation—The actual cost of using a subcontractor is the cost of the work plus the cost of the workers' compensation insurance whether you pay for it in the subcontract or separately. When negotiating a subcontract, you should establish a clear understanding with the subcontractor about workers' compensation insurance and write it into the subcontract. You need to establish whether or not the subcontractor has his or her own coverage and, if so, when the subcontractor will deliver the certificates of insurance to you so that you will not have to pay for the coverage too. Be sure you have a system to alert you if a subcontractor's current certificate is not on file. Some builders, remodelers, and developers will not do business with a subcontractor who cannot offer proof of being fully insured. If you do use a subcontractor who does not have workers' compensation coverage, spell out in the subcontract whether or not you will withhold the cost of the premiums from your payment.

Monitor Subcontractor Schedules

Coordinating the schedules of your subcontractors strongly affects how well you control costs and thereby increase profits. This coordination is primarily a case of the builder, remodeler, or developer having a businesslike attitude toward the jobs. The skills required to schedule the subcontractors are explained in detail in *Scheduling for Builders.*[9] For example, if you have several jobs in progress, you may have to compare the float period of one job with the time slot for similar work on another job to effectively direct where and when your subcontractors work. You also need to consider job cleanup, preparation for inspections, scheduling, waiting, and fixing any deficiencies in your agreements with subcontractors.

Control Payment Process

Specify Details of Payment—Make sure that your subcontract is specific about how payment will be made. For example, if a subcontractor's work will be spread out continuously over a period of weeks, you may need to make progress payments. You should establish and follow a policy that you pay a subcontractor for work only after you or someone on your staff has inspected that work and verified that it was done satisfactorily.

Human nature causes people to focus their attention where they have the potential to achieve the greatest reward. If a subcontractor has been paid for work not done because nobody verified his or her request for payment, he or she may focus on other work that adds income instead of work for which you have already paid. Many builders, remodelers, and developers pay subcontractors every Friday for work they have completed through Thursday afternoon. This practice gives the builder or remodeler Friday morning to inspect the subcontractor's work for which payment is requested.

Prompt Payment—You should not try to make a subcontractor finance work he or she has done by withholding payment after it is due. Rapport with your subcontractors has value, and one way to establish a strong working relationship is to make payment when payment is due.

Handle Backcharges, Call-Backs, and Warranty Work

You need a standard procedure for subcontractors regarding backcharges, call-backs, or warranty work. Backcharges can result from a subcontractor damaging work installed by others or from your having to provide labor, materials, or equipment to do a part of a subcontractor's scope of work.

For example, suppose the trim carpenter drives a finish nail into a water supply line and the water leak damages the drywall. Customarily the builder or remodeler would withhold the amount required to repair the plumbing, the drywall, and any other work damaged by the leak. Losses to the builder or remodeler occur when the source of the damage is impossible to determine. Builders and remodelers often have to absorb the cost of broken windows, cigarette burns, and similar accidents. Each subcontractor should deliver and warrant high-quality work and the material provided.

Scheduling warranty work is important because the gravity of a problem tends to increase in the owner's mind when nobody appears to be doing anything about it. You should schedule all warranty work as soon as possible after the homeowner reports the problem.

Employees

An employee differs from a subcontractor in that he or she is salaried or works by the hour and is directed to do tasks on a continuous basis. Of course a subcontractor may also be paid by the hour. For instance, grading contractors grade lots at an hourly rate. However the main difference between dealing with subcontractors and regular employees is that you do not have to pay payroll taxes and social security benefits on subcontracts. But the lines between employee and subcontractor are not clear cut. If you have any questions regarding whether or not those who work for you should be treated as employees or subcontractors, you should get an opinion from an attorney. If you do not comply with applicable regulations, the fines, back taxes, and interest, could cost you your business.

Hire Employees

The process of using direct hourly labor for certain tasks begins with the decision that hourly labor is better for that task than subcontract labor. Today in most locations in the United States, you can find subcontractors to do virtually any task required in homebuilding. Remodelers may find that hourly labor is better for them for certain tasks because hidden conditions make the scope of the work uncertain.

In small organizations usually the principal person (the owner or manager) does the hiring. In large companies superintendents or others may hire hourly labor. Regardless of the size of your business, you need to develop written personnel policies because many regulations imposed by different government agencies affect what you do in hiring, supervising, and firing people. You must be sure to comply with these legal requirements or risk heavy fines. Some of these are discussed in the paragraphs below.

The federal government and some state and local jurisdictions have laws against asking questions about an employee's marital or family status, sex, age, dependents, birth control, pregnancy, disabilities, place of origin, and relatives. If you have any questions about these issues or interviewing, hiring, and firing, you should consult an attorney. Violating any of these laws can prove costly.

You need to verify that potential employees are either citizens or entitled to work under the applicable federal, state, and local laws. You should have the job candidate fill out a federal W-9 form. If you have any questions about your obligations regarding a potential employee's citizenship, you should contact the U.S. Immigration and Naturalization Service for complete instructions. After an employee is hired, you need to obtain the person's complete name, address, and social security number. The employee must fill out and sign a Withholding Allowance Certificate or W-4 form—and sometimes forms for state and local income taxes—to allow you to withhold the employee's taxes.

Employees may be found in any number of ways. Current employees or subcontractors may know of qualified individuals looking for work. Previous employees are often a good source of potential workers. Employment agencies, newspaper advertising, union halls, and state unemployment offices also may be possible sources. Hiring and firing practices directly affect cost control. For example, hiring people well-suited to the job not only increases quality but also keeps turnover costs low.

Train Employees

Training craftspeople to replace those workers who retire or otherwise leave the workplace is a problem of major proportion. In residential construction, where most of the work is done by subcontractors, the problem for builders and remodelers is to find subcontractors who have qualified people. Certain trades (such as electrical and plumbing) have a built-in quality control mechanism through licensing requirements while some other crafts have little or no control or training programs. The Construction Industry Institute has tracked a steady decline in construction productivity rates over the past decade.[10] A single builder or remodeler has difficulty making an impact on this critical problem. However, if the awareness of the problem is heightened, those people affected can work through local home builder associations to improve the situation through such programs as the Job Corps training sponsored by Home Builders Institute. In instances where builders and remodelers contribute to employee training, a fairly common result is improved quality and productivity, a reduction in accidents, and lower costs.

Keep Time Cards

Timekeeping has two basic functions: One is to determine how much to pay the employee; the other is to determine to which work categories to charge the wages.

Hourly employees will always have someone to whom they report directly. This person may be the builder, remodeler, developer, superintendent, production manager, or lead carpenter. This person should keep a time book to record the hours and how the time is to be charged. Transferring the time data to time cards or time sheets on a daily basis is a good idea. You should keep the time cards or time sheets at the office because they will become the basis for making up the payroll.

After gathering and checking the time sheets, the bookkeeper figures the gross pay and deductions and checks that all numbers are transferred and added accurately. He or she also calculates and records the cost of the employer's payroll taxes and workers' compensation insurance on the appropriate forms. Lastly you need to charge all of these costs to the individual jobs that were worked on that week. In small-volume businesses the bookkeeper could be the owner, an employee, an employee of a bookkeeping firm, or a freelance bookkeeper. This time-sheet information is valuable to builders, remodelers, and developers in planning future jobs. Remodelers can use it to make up their own cost books, but remodelers costs may vary from job to job considerably because they often have to (a) perform labor-intensive tasks and (b) match and use materials that are hard to work with.

No matter how small your company is, you may want to have a payroll bank account that is separate and apart from your general account.

You should deposit all revenues to the general account, and as funds are required for payroll, write checks from the general account to the payroll account. In this way you can summarize all the individual payroll checks in the general account as one entry instead of one by one and record their expense easily on the job ledgers. Therefore you can post your ledgers from the general account alone, and the payroll account becomes a subsidiary account of the general account.

In addition to giving the employee his or her check, you must also provide a stub or memorandum itemizing the deductions. Furthermore, before January 31 of each year, you must give the employee a Withholding Tax Statement, Form W-2, showing the total earnings and tax withholdings for the year. At the end of each quarter, you need to reconcile social security and withholding tax returns and the employee's share of the contribution required under the Federal Insurance Contribution Act (FICA) with the payroll records on Form 941 and confirm that all deposits are up to date.

Many commercially available computer software programs are designed to perform payroll functions and job cost postings. Most local computer software stores are equipped to assist you in finding the best system to suit your needs. Having payroll and job cost modules is a labor saver for the office. But talk to other builders, remodelers, and developers before purchasing anything and get the benefit of their experience. (See Software Review Program in Chapter 3.)

If the payroll account is done exactly right, it will go to zero when you have made the quarterly payments. However you should consider keeping a cushion of money in the account to prevent the bank from charging you service fees. You also need to distribute all federal earnings statements (Form 1099) to subcontractors and others to whom you owe them by January 31 of each year.

On occasion an employee may be unable to pick up his or her paycheck. If this happens, you should be extremely careful in turning the check over to a third party. In the absence of a written authorization signed by the employee, you may be responsible if the employee does not get his or her money.

Occasionally you may be confronted with a legal attachment to an employee's pay. Exspouses and creditors frequently attach or garnish an employee's wages. You need to be careful in these cases because you may become personally liable if you try to circumvent a valid claim and pay the employee despite the attachment.

Well-trained, loyal employees can form the basis of a stable organization and a strong line of defense against run-away costs. You should use every effort to foster an atmosphere of teamwork and unity. This effort begins with an understanding on the part of supervisors that employees are human beings and deserve the respect that the supervisors want for themselves.

You should promote the concept of teamwork and involve employees in problem solving and policy making as part of your cost control strategies. If your employees and subcontractors believe they are part of the problem-solving team and that their suggestions regarding policy decisions will be considered, your overall quality and productivity probably will increase. Likewise if these people are aware of cost control policies and measures and asked to help solve cost problems, they will be more likely to respond positively and to help keep costs down.

Equipment

Renting Versus Owning

You can rent virtually any piece of equipment that you might need from an equipment-rental company. Obviously you are better off to rent equipment that you seldom use rather than buying it. For most builders, remodelers, and developers, the decision to buy or lease is an economic one. If owning a piece of equipment is more profitable than renting, you should buy. Maintenance and upkeep should be included in the analysis of whether to rent, lease, or buy. You must consider the answers to such questions as those listed in Figure 5-4 in deciding whether to buy, rent, or lease equipment and discuss the tax aspect of each possibility with your tax accountant.

Repair and Maintenance

Maintenance costs represent a variable expense that can make ownership undesirable. Repair and maintenance expenses on some types of equipment make up a large proportion of the cost of ownership. On these items, if you do not have in-house repair capability, renting or leasing is often better.

Figure 5-4. Questions to Ask Yourself About Renting, Leasing, or Buying Equipment

- Do you have enough work for the equipment that the rent you pay would be more than enough to buy the equipment and cover interest, maintenance, storage, trailering, insurance, and repairs?
- Do employees need special training to use the equipment?
- Is that employee's workload already too heavy?
- Are the necessary skills easy to learn?
- Does the equipment require highly skilled maintenance service and repairs or could an employee handle the service and repairs?
- Will the equipment wear out easily and need to be replaced frequently?
- How frequently is the equipment likely to need such service and repair?
- Is that employee's workload already too heavy?
- Do you already have a safe place to store the equipment when it is not in use?

Depreciation

Accounting for the costs of owned equipment is relatively simple. Builders, remodelers, and developers customarily charge costs for any company-owned equipment that can be expensed out for tax purposes in the first year of ownership to the general overhead account. These costs are then absorbed by the individual jobs on a pro rata share basis. Because depreciation expenses for equipment is fixed and recurring in nature, a common practice is to charge these expenses to the general overhead account.

Another way to account for depreciation on equipment is to set up a separate internal account for equipment operation. This account is treated like a separate profit center. The expenses charged to this profit center are the costs of repairs, parts, interest on invested capital, depreciation, and other associated costs that may be expensed. The account would be credited when equipment is rented back to the company and that cost charged to individual jobs.

Rental Procedures and Keeping Time Records

If you choose to set up a separate profit center for equipment, you should charge the established internal rental rate for each piece of equipment to your various jobs. When you use the equipment on a job, keep its time the same way as you would for an employee. Record the number of hours each piece of equipment is used for a given job, and transfer this information to an equipment report that includes—

- the name and identifying number of the piece of equipment
- the job name and/or number on which it was used
- the date
- the amount of usage

The usage fee becomes the income for the equipment operation account. Charge it to the appropriate job ledger just as if you had rented the equipment from an equipment-rental company. At the end of the year, any profits earned by the equipment operation account become profits to the company, and any losses are charged to general overhead.

Customer-Related Costs

On contract work or on speculative projects that sell before completion of the work, you must deal with the owner on several issues. You need a thorough understanding of the effects of your decisions to keep customer-related costs under control.

Change Orders

Changes in the scope of the work sometimes cause misunderstandings between builders and remodelers and their clients. Change orders are a normal part of construction work. However, unless a change order clause in the contract allows the owner to make changes, the owner does not have an automatic right to do so. Only in extreme cases, such as when the owner is totally unreasonable, would you refuse to make a reasonable change. Usually changes to the scope of the work result from design changes or accommodations to site conditions different from those that were expected, such as encountering more rock than anticipated, unforeseen mechanicals (plumbing or electrical) in the walls, and unexpected framing requirements. To avoid losing money, the builder, remodeler, and developer must understand all the potential costs involved in change orders.

Basic Elements to Track for Cost Control

The builder, remodeler, or developer and owner must realize that the job, as originally planned, will be disrupted by changes and will usually cost more. When changes are made to a project, schedules change, materials change, delivery dates change, and methodology changes. These disruptions may be minor or major. Minor changes can sometimes have major repercussions. In your markup you should recover the cost of the extra management effort needed to make these changes. For this reason your markup on change orders has to be greater than the markup on the base work.

Job Overhead—Certain aspects of job overhead are directly related to time. If a change order extends the time of the project, the builder or remodeler must be sure to properly consider how job overhead will increase as a result. Even a change that costs no more in labor or materials will cost you more if it extends the construction time. Interest expenses, tying up equipment the superintendent needed elsewhere, and other time-related costs can make job overhead charges higher. You may need to price the job overhead portion of a change order on how much extra time it takes in addition to how much more it costs.

General Overhead—If a change order or some other factor causes an extension of the contract time, the office support for the job continues longer than originally planned. Many builders, remodelers, and developers cover general overhead expenses on projects by adding a percentage increase appropriate for bidding the base contract (see Chapter 4). However this method does not cover the costs incurred if the job lasts longer than expected without bringing in any extra income.

For example, assume that you are doing a $200,000 job and you figure your general overhead to be 6.2 percent. You would charge $12,400 of your general overhead expenses to this job. Assume further that the job was planned to take 6 months but, because of changes in the work that did not increase the direct job cost, the job lasted an additional 3 months. During that 3-month period, your office would have to continue to support the job and, therefore, be unable to support other work. For this reason, you need to charge a price that covers the extra time spent on the job as well as the costs added to the job.

To charge the additional overhead properly you need to know how much general overhead a job has to absorb every day. Given the example described above of a job lasting 6 months with a general overhead of $12,400, if you have 5 holidays per year leaving 255 working days per year or 128 days for 6 months, the job has to absorb $12,400 divided by 128 days or $97 per day in general overhead expenses. To be more accurate, you would calculate the daily general overhead charge and multiply that by the anticipated extra work days rather than simply multiplying the cost increase by a percentage.

Customer Service

Maintaining good customer relations sustains a builder's or remodeler's reputation. The way you handle changes directly affects this relationship. You should explain to the owner at the time of contract (and in writing) that to give the buyer or owner the best price possible you have included in your price only the items specified and shown in the drawings. Furthermore, you have not charged for anything not included in the contract or shown in the plans and specifications. For this reason, if the owner wants any extras, you will have to charge a fair price.

Many owners believe that they should be able to add all the extras they want at no extra cost. For the sake of customer relations, some builders do not charge for small extras. If you choose to follow this practice, you need to be aware of exactly what you are doing and of how much it is costing you. On cost-plus work you will automatically be paid for all extras. But you need to keep up with changes and their costs. A strong legal precedence maintains that a builder or remodeler should keep the cost-plus buyer or owner up to date on the costs of changes and the final estimated cost throughout the job. You should document all changes completely with copies to all parties, whether on a fixed-price or cost-plus project.

Processing Change Orders

If a change involves an extra expense, you should always get the owner to sign the change order before proceeding with the work. You should not trust your or the owner's memory by entering into oral transactions, especially where money is involved. Because memories fade, you risk losing money or the owner's goodwill if you do not write down and both sign and date the details of any agreements.

Punchlist and Warranty Work

Handling punchlist work and warranty work in a proper manner can enhance a builder's or remodeler's reputation and produce additional and more rewarding work in the long run. Conversely not doing a proper job of punchlist and warranty work can hurt you. Handling punchlist work and warranty work can present special problems in tracking costs. At the end of a job during a walk-through, the owner and the builder or remodeler compile the punchlist (the items that still need to be done). Warranty work includes items that need to be fixed or corrected that are discovered after final completion and during the warranty period. Subcontractors should do most of the punchlist and warranty items for the items subcontracted to them. However for various reasons, the builder or remodeler may have to pay for some of the punchlist and warranty work.

As a public relations policy, some builders and remodelers contact the owner shortly before the end of the warranty period to see if everything is in order. In a business in which reputation is of paramount importance, such policies will put you ahead of the your competition.

Other Contingencies

You should close the books on a job as soon as the punchlist work is completed. Of course, if anyone finds and reports any warranty work before closeout, you should charge doing this work to the job ledgers. You can charge the cost of warranty work after the books are closed to the general overhead ledger by setting up a warranty ledger that is subsidiary to the general ledger (to see which job cost codes are causing you the most trouble). Or you can reopen the books and amend the job ledger.

Every builder, remodeler, and developer should establish procedures to handle purchasing, subcontracting, the hiring and overseeing of employees, the rental or ownership of equipment, and the handling of customer-related costs. Such practices are at the heart of effective cost control.

Six

Track and Analyze Remodeling Costs

Remodeling an existing home can be more challenging than building a new home from scratch from a management viewpoint as well as from a technical standpoint. In trying to adapt or preserve an older dwelling, the remodeler must be ready for surprises and tough problems. To succeed, the remodeler must have, not only a thorough knowledge of materials and construction techniques used in earlier times, but also the skills required to track and account for unforeseeable costs.

Budget for a Cost-Plus or a Fixed-Price Contract

Although whether a remodeling job is priced on a cost-plus or fixed-price basis may be vitally important from other standpoints, insofar as tracking job costs, it makes little difference. Both kinds of jobs start out with a budget for individual work categories. You base this budget on an estimate of your understanding of the scope of the work. In this regard you must spell out in the contract all assumptions you made about the scope of the work as you estimated the job for both fixed-price and cost-plus work. In addition you need to say what is not included in the scope of the work that a customer might think is included.

For example, the contract should especially mention and omit from the scope of work any improvements that your customer considered earlier but later decided against. In the process of doing the work, every time you encounter a condition different from what you assumed, you need to document the differing condition, inform the owner in writing of the differing condition, and get a signed change order prior to expending any money on the change.

Change Orders

Change orders are typically handled in one of two ways. The extra cost may be paid for on either a fixed-price or a cost-plus basis. On fixed-price projects, change orders may be handled either way. On projects that are paid for on a cost-plus basis, typically you would also handle the changes on a cost-plus basis. On cost-plus work you need to track and account for the costs associated with performing change-order work including unforeseen conditions separately.

For example, you may have a job that is cost plus 30 percent to cover profit and general overhead. Further assume that you estimated the cost to be $87,500. In the middle of the demolition work, you discover some substandard structural lumber. You should immediately notify the owner of the condition and request permission in writing to extend the demolition work to make sure you have discovered the extent of the problem.

At this point you may or may not be able to estimate the cost of the extra demolition with reasonable accuracy. If you can, you should inform the owner (a) how much more the demolition work is likely to cost and (b) upon completion of this effort, you will be able to estimate the cost of the extra materials and labor to replace the substandard framing.

Prepare the Change Order and a Cover Letter

Assuming the owner gives you the go ahead, you should immediately prepare a change order memorializing your conversation. To avoid delays, you may want to hand deliver the change order. In the change order you should refer to the condition, the estimated increase, the new anticipated completion date, and the approval to proceed. After you receive the signed change order and finish the demolition, you should estimate the cost to replace the substandard lumber and notify the owner in writing. Your cover letter for the extra exploratory work might read something like the one in Figure 6-1. A sample change order for extra exploratory work appears in Figure 6-2.

Figure 6-1. Sample Cover Letter for Change Order for Extra Exploratory Work

Dear ______________________:

As I explained during our conversation today, our carpenters have discovered that the wall between the unfinished basement and the game room is infested with termites. We estimate that an additional $500 will cover the cost of the extra demolition. This portion of the extra work associated with the substandard lumber will increase the budget from $87,500 to $88,000. We will proceed with the work on a cost plus thirty percent basis as soon as we have your signature on the attached change order. We will contact you immediately upon completion of this extra demolition to discuss options available to us.

Sincerely,

____

President

Figure 6-2. Sample Change Order for Extra Exploratory Work

CONTRACT CHANGE ORDER

CENTURY CONSTRUCTION COMPANY
Post Office Box 81743
Atlanta, Georgia

Project: Lambert Residence
345 Dean's Trail
Roswell, GA

Change Order Number 1
Date: August 17, 1995

Revised Contract Amount

Previous Contract Amount: $ 87,500.00
Estimated Cost of Change: + 500.00
Estimated Revised Amount: $ 88,000.00

An (increase) (~~decrease~~) (~~no Change~~) or 2 days in the contract time is hereby authorized.

This order covers the extra work as follows:

Extra demolition of frame wall between unfinished basement and game room to discover extent of termite infestation.

Basis of Payment: (Cost Plus) (~~Lump Sum~~)

Note: If the change is to be paid on a cost plus basis the cost portion shall include the cost of material, labor, subcontractors, equipment, and additional cost of supervision and field office expenses attributable to the change plus 30 percent

The work done as a part of this change order shall be done in accordance with the same terms and conditions as the original contract.

APPROVED:

____________________ (Owner)

____________________ (Contractor)

Now assume that, after you have completed the demolition and had a pest control company assess the damage, you determine that the defective material needs to be replaced and the structure treated to prevent further deterioration. After you have estimated the extra costs, you should inform the owner in writing. Assuming the owner grants approval, your follow-up letter might read like the one in Figure 6-3. The change order for the extra work would look similar to the one in Figure 6-2 for the additional exploratory work.

If the work that needs to be done involves lead abatement, asbestos removal, or anything in which you have no expertise, you should suggest that the work be done as a separate contract between the homeowner and that firm. Such work is often regulated by state, local, or federal law. Without proper training and certification or licensing, a remodeler could be violating a law and become exposed to unwanted liability from the Environmental Protection Agency, the Office of Safety and Health Administration, and the homeowners.

Change Orders and Variances

If you have established the budget for each category and the scope of the work for a category changes, your actual costs for that category will automatically be different from the original budget. If you do not change your budget for a category affected by a change, the entire effect of the change will show up in the variance, that is, the difference between your budget and the actual expenditure. In analyzing the variance you will need to separate out the costs associated with the change order from those resulting from other causes.

Figure 6-3. Sample Cover Letter for Change Order for Extra Work

Dear ___________________:

In accordance with our phone conversation today, our estimate for the extra structural framing work at your home necessitated by defective lumber is $4,250. Additionally, we recommend that we contract with a competent pest control company to treat your residence and provide an appropriate termite bond. We recommend that you undertake this effort as a change order to our contract so that we can schedule the work and check to be sure it is done properly. The Termired Corporation has inspected the property and has offered to perform this task for $715. (See attached quote.) This addition of $4,250 for the carpentry work and $715 for the termite treatment brings the revised budget to $92,965. We will proceed with the work as soon as we have your signature on the attached change order.

Sincerely,

____

President

Accumulating costs of changes in a category without reflecting a change in that category's budget during the job and before the final accounting is made, prevents you from making accurate midcourse assessments without remembering and calculating each change and its dollar value and recording it to the proper budget category.

Extra Costs Related to Change Orders

To consider the effect of the changes during construction, you can account for extra costs related to change orders in posting your job ledgers in two ways. You can estimate each and every change using all the same budget categories that were used in the base contract estimate and adjust the budgeted amount for each category, or you can create a separate budget category solely reserved for changes.

If you use a separate budget category, you would start out with a budgeted amount of zero for changes. In this way the budgeted amount for each category remains the same, and the variances represent differences other than approved changes.

If you adjust each budget category as the changes are approved, you can track and report all the costs for each category without regard to whether or not the costs are for base contract work or extra work. However, if you choose this method, you must separate the items in your estimate of the change order into the individual categories.

If you choose to account for all costs for changes in a separate changes category, you must separate out costs incurred for change orders from costs for base contract work. On the other hand, if you establish a separate category, you must track billings for extras and differentiate them from base contract prices.

For example, assume that your budget for framing is $14,400 and your budget for roofing is $2,360. The owner wants to make the front porch larger than planned. You estimate that the framing cost will increase $920, and the roofing will cost $110 more. If you choose to increase the budget for each category, your revised budget for framing would be $15,320 and the revised budget for roofing would be $2,470. As your subcontractors or employees perform the work, you would want to know the total cost of all the various elements involved in the framing and for those in the roofing. You would compare these costs to the revised budget.

If you choose to account for all changes in the work under a separate category, your budgets for framing and for roofing would remain the same, and you would add $920 plus $110 to the changes budget. You would notify your person in charge in the field to keep a record of the costs for the changes separately. When you have obtained these separate costs, you would post them under the changes category.

In setting up job ledgers for remodeling projects, you can use the categories established earlier for builders by making a few modifications.

Several items of work that are common to remodeling do not often occur in new building projects. This work includes such tasks as—

- demolition
- moving building segments such as windows, doors, wiring, and plumbing
- tying into the structure
- floor leveling
- protection of the homeowner's property
- daily cleanup
- matching and acquiring hard-to-find materials

Your decision on how to charge the various work items may depend on the kind of work in which you ordinarily engage. In making this decision, you usually need to keep the work done by a single subcontractor in the same category. Practically, if you are making payment to a subcontractor, that subcontractor's work will usually fall under one category, unless the subcontractor is doing work on more than one clearly different task. In either case, have the subcontractor break down the invoice in the same way your budgeted categories are set up.

Determine Profits

The basic rule for determining how much profit to add to a job is to add the amount you reasonably can and still get the job. Sometimes you have competition, and your price needs to be competitive in order to get the work. For negotiated work you are restrained by your client's budget, by industry customs and practice regarding how much profit the kind of work you are furnishing is usually marked up or by your conscience.

People make money by working by the hour, by taking a risk, or by a combination of the two. The risks for a builder who does little or none of the work with his own labor is not as great as for a subcontractor or a builder who does the work with his or her own labor force. Usually the biggest variable in the cost of a project is the labor cost.

If you have a contract to furnish $100,000 worth of materials only, and all you have to do is to order the materials, bill for it, and pay your suppliers, you would probably be willing to take less profit than if you had to furnish $100,000 worth of labor.

Generally subcontractors and remodelers whose work consists of providing a large percentage of labor need to mark up their costs more than a builder who subcontracts everything because subcontractors are dealing with a substantially smaller amount of money for a considerably larger effort.

In addition to the risk factors, a builder subcontracting a majority of his or her work can produce a much greater volume than a subcontractor or remodeler who expends the same effort.

Profits for change orders should be higher than for base contract work.

Changes in the scope of work mean changes in material orders, scheduling, and sequencing of activities. When you establish your criteria for building the base contract, you establish a list of materials, a schedule, and a methodology. This takes effort. When you change the scope, you may have to undo your previous efforts, and therefore you should charge more profit to compensate yourself.

Allocate Expenses and Track Costs

Some of the work items may seem to fit into more than one category. For example, if a project has demolition of framing work or floor leveling, these items may be classified as site work since you are preparing the "site" to do your work. However, if this work is done by carpenters, you may want to allocate these expenses to framing or carpentry. You would especially want to follow this practice if you are tying your workers' compensation classifications into your job costs via the payroll module where job cost categories must fit insurance classifications accurately. If you allocate carpenters to the sitework category, you would pay a different (possibly higher) rate for workers' compensation insurance.

As an example of how a remodeler might track the costs associated with his or her work, consider the following project. An existing one story house has a 20- by 30-foot screened porch. The floor is a concrete slab on grade that is approximately 3 inches lower than the slab level of the house. The roof structure of the porch is corrugated fiberglass on 2x6 rafters. The rafters rest on beams that in turn are supported by wood posts. Three pairs of French doors open from the house to the porch. The scope of the work includes the activities in Figure 6-4, and a recap of the estimate appears in Figure 6-5.

Figure 6-6 shows a job ledger sheet that reflects the costs for the job. The budget for each category, as established from the estimate, appears above each category. As you or your bookkeeper write checks or establish that

Figure 6-4. Activities in the Scope of the Work

- Convert the porch to a den.
- Add a second floor over the new den for two bedrooms and two baths creating a total of 1,250 square feet of new living space.
- Add a side porch next to the new den.
- Add a balcony for the new bedrooms over the side porch.
- Add a new 11x15-foot front porch with roof.
- Disconnect and raise the existing garage roof 3 feet to match the new design.
- Replace the water heater and move it to different location.
- Remodel the master bath.
- Replace three windows.
- Remove and dispose of the existing French doors to the existing porch.
- Remodel the kitchen.

Figure 6-5. Recap of Estimate

Job overhead	$2,250
Sitework	1,400
Footings and slabs	1,590
Masonry	895
Framing	14,400
Roofing	2,360
Plumbing	3,800
Electrical	3,600
HVAC	2,540
Insulation	710
Drywall	1,400
Interior trim	4,630
Paint and wallcovering	3,000
Floor covering	2,770
Cabinets	5,500
Appliances	1,990
Exterior	500
Total	**$53,335**

debts are owed, you should post them in the ledger. The ledger includes a separate category for change orders with a budget value of zero. The remodeler or the owner might like to budget some money for changes, but to accurately estimate their final cost is extremely difficult if not impossible and interferes with meaningful comparisons.

If the budget for change orders for this job had been set at $10,000, the total job budget would be $63,335 instead of $53,335. Since the actual total cost was $56,486.88, establishing the changes budget at $10,000 would seem to indicate that the job came in $6,848.12 under budget. However, a more accurate statement would be that the job actually came in $3,151.88 over budget and that $2,575.40 was in changes. This calculation leaves a total of $3,151.88 less $2,575.40 or $576.48 as a variance from the estimated budget that needs to be analyzed.

If the system is set up in this manner, the variance in each category is analyzed independently of the change orders. You can choose not to have a category for approved changes and post the costs for the changes in the same categories as for the base contract work. The individual changes will become an element of your variance analysis for these categories and require more work to accurately analyze each category. The separation of change orders also allows them to be treated as a separate profit center. Thus you can check invoices against estimates of change orders.

In remodeling, change orders are either customer driven or situation driven. Either the customers change their minds about something or a situation discovered requires a change. The situational changes often must be done quickly because taking time to get estimates will throw the job schedule off or cause other undue disruption, so many remodelers provide ballpark estimates for the change and charge on a cost-plus basis. The idea is to be sure the customers pay for the work they authorize not to penalize them. A checklist for tracking and analyzing remodeling costs appears in Figure 6-7.

Figure 6-7. Tracking and Analyzing Remodeling Costs

- Make sure your agreement with the owner covers all pertinent considerations.
- Keep the owner appraised of changes in the work.
- Develop a cost-category system that is tailored to your needs.
- Establish a budget for each cost category.
- Constantly compare costs with your budget.

National®Brand 45-621 Eye-Ease® Made in USA

Lambert Residence

	Initials	Date
Prepared By		
Approved By		

date		Payee	Ch #	Amount (1)	Subtotal (2)	Job overhd (3)	Site work (4)	Ftgs & slabs (5)
				Budget	53,335	2,250	1,400	1,590
6	12	City of Gotham	428	450 -	450 -	450 -		
6	19	Roy's Carpentry Serv.	437	850 -	1300 -		850 -	
6	20	Slab Prep, Inc.	440	360 -	1660 -			360 -
6	22	Footing Diggers, Inc	460	50 -	1710 -			50 -
6	24	Joe's Backhoe Service	462	647.20	2357.20			
6	27	Concrete Finishers, Inc	468	160 -	2517.20			160 -
6	27	Termirid	469	200 -	2717.20			200 -
7	1	Brown Trucking	475	80 -	2797.20			
7	10	Bluff River Readimix	485	561.20	3358.40			561.20
7	10	Fixit Masonry Co.	486	900 -	4258.40			
7	10	Home Lumber Co.	488	1010.60	5269.00			271.40
7	10	Apex Brick Co.	489	748.20	6017.20			
7	15	Roy's Carpentry Service	501	5200 -	11217.20			
7	20	D & B Plumbing	515	2100 -	13317.20			
7	23	Dynelectric	517	1600 -	14917.20			
7	24	AAA Heating & Air	520	1270 -	16187.20			
7	27	Blue Ridge Insulation	522	695 -	16882.20			
7	31	Gulf States Roofing Co.	530	998 -	17880.20			
8	3	Drywall Hangers, Inc,	535	400 -	18280.20			
8	6	Drywall Finishers Inc.	537	640 -	18920.20			
8	10	Roy's Carpentry Service	540	3250 -	22170.20		550 -	
8	10	Home Lumber Co.	541	8774.53	30944.73			
8	20	Larimat Cabinets	560	5500 -	36444.73			
8	24	Roy's Carpentry Service	569	2700 -	39144.73			
8	31	Donald Ray's Paint Co	571	3000 -	42144.73			
9	5	D & B Plumbing	592	1900 -	44044.73			
9	8	Dynelectric	600	1600 -	45644.73			
9	8	AAA Heating & Air	601	1270 -	46914.73			
9	10	Cox Electric Supply	611	398.27	47313.00			
9	10	Ace Appliance Co	612	2061.14	49374.14			
9	10	Home Lumber Co.	613	2514.12	51888.26			
9	12	Bob's Cleaning Service	615	100 -	51988.26			
9	15	Dalton's Carpet Co.	618	2806.50	54794.76			
9	20	Payroll Account	628	1692.12	56486.88	1692.12		
		ACTUAL EXPENDITURES			56486.88	2142.12	1400 -	1602.60
		VARIANCE			<3151.88>	107.88	—	<12.60>

Figure 6-6. Sample Job Ledger

6	7	8	9	10	11	12	13
895	14,400	2,360	3,800	3,600	2,540	710	1,400
Masonry	Framing	Roofing	Plumbing	Electrical	HVAC	Insulation	Drywall
739.20							
	5000.–						
			1900.–				
				1600.–			
					1270.–		
						695.–	
		998.–					
							400.–
							640.–
	2600.–						
	6990.27	1360.10					424.16
			1900.–				
				1600.–			
					1270.–		
				398.27			
739.20	14590.27	2358.10	3800.–	3598.27	2540.–	695.–	1464.16
155.80	<190.27>	1.90	–	1.73	–	15.–	<64.16>

Figure 6-6. Sample Job Ledger (Continued)

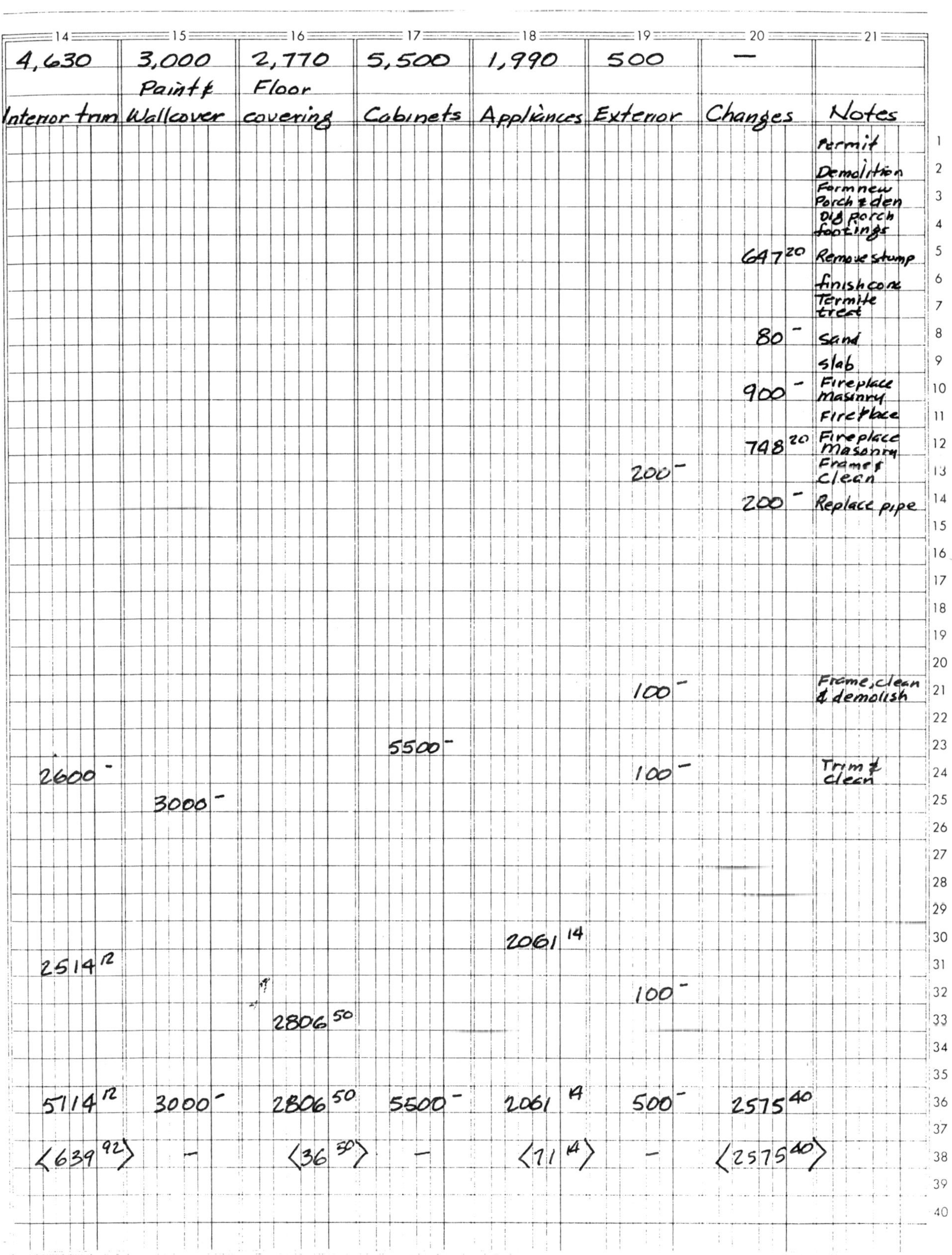

	14	15	16	17	18	19	20	21
	4,630	3,000	2,770	5,500	1,990	500	—	
	Interior trim	Paint & Wallcover	Floor covering	Cabinets	Appliances	Exterior	Changes	Notes
1								Permit
2								Demolition
3								Form new Porch & den
4								Dig porch footings
5							647^{20}	Remove stump
6								finish conc
7								Termite treat
8							80 -	Sand
9								slab
10							900 -	Fireplace masonry
11								Fireplace
12							748^{20}	Fireplace masonry
13						200 -		Frame & clean
14							200 -	Replace pipe
15								
16								
17								
18								
19								
20								
21						100 -		Frame, clean & demolish
22								
23				5500 -				
24	2600 -					100 -		Trim & clean
25		3000 -						
26								
27								
28								
29								
30					2061^{14}			
31	2514^{12}							
32						100 -		
33			2806^{50}					
34								
35								
36	5114^{12}	3000 -	2806^{50}	5500 -	2061^{14}	500 -	2575^{40}	
37								
38	⟨639^{92}⟩	–	⟨36^{50}⟩	–	⟨71^{14}⟩	–	⟨2575^{40}⟩	
39								
40								

Figure 6-6. Sample Job Ledger (Continued)

Seven

Track and Analyze Development Costs

The building business is compartmentalized into many interconnected segments. For example when a buyer purchases a home, the businesses that may realize profits from the planning stage to move-in include some or all of those listed under Firms Unusually Involved in Figure 7-1. This figure also lists business services and other construction-related specialists that might be involved in the development, building, and sales loop of creating a house for a customer. These firms usually profit from their activities that help to create new housing.

The developer purchases the land; performs the necessary land planning, site design, and regulatory compliance functions; installs the improvements; and sells the individual lots for a profit. Banks lend money to the developer, the builder, and others and earn income from commitment fees, discount points, and interest. The builder organizes the materials, the manpower, and equipment to construct the home and earns a profit from these activities and building the house. The subcontractors maintain and oversee a workforce, often furnish materials, and mark up their costs to realize a profit. Material suppliers establish wholesale accounts, maintain an inventory, make jobsite deliveries, and earn a profit from markup on these materials and services. Equipment rental companies purchase and maintain a variety of equipment and profit from their use. Realtors become licensed, show homes to prospective buyers, and earn a commission.

The various participants in this process may perform all or part of the tasks of other participants to increase profits. As an example, a builder may develop a piece of property, finance the development and the house construction from his or her own pocket, do the landscaping with his or her own hourly labor, and use the company-owned front-end loader to do the rough site grading. In this situation the builder usually makes a profit on the building of the home, the land, the landscape subcontract, and one piece of equipment. The builder also might realize a savings on the construction and development financing.

Figure 7-1. Businesses That Profit from Land Development, Construction, and Sale of a House

Firms Usually Involved

- marketing person, firm, or consultant
- Realtors (for the sale of the land and the sale of the house)
- land development firm
- sewer and other utility installation companies
- engineer
- banks
- insurance broker
- building company
- subcontractors (including excavators; masons; framers; heating, air-conditioning, and ventilation specialists; plumbing; electrical; insulators, landscaping, painting and wall-covering)
- material suppliers
- equipment rental companies

Other Construction-Related Specialists

- environmental impact and/or hazardous waste specialist
- equipment dealers
- interior decorator (if house served as a model)
- product showrooms and furniture dealers
- sod companies
- surveyors

Business Services

- answering service
- bookkeeping, accounting service, and/or payroll service
- freelance writers, artists, and others involved in promotion of marketing and publicity materials
- printers of marketing and publicity materials

To successfully compete in land development, a person has to be able to estimate all the costs accurately or have it done by a consulting engineer. A land developer must be able to pick the right piece of land to develop at the right price. He or she also must be able to reasonably predict the economic future of the local target market based on careful study of that market. Developers get hurt financially when they get caught with an inordinate number of lots in inventory during downturns in the economy and have difficulty meeting their financial commitments.

Developers are particularly vulnerable to downturns in the market that slow down construction because they almost exclusively develop residential lots on a speculative basis. Sometimes development projects take a long time to sell out. Predicting the future of new home sales is extremely difficult. Generally, when the interest rates are low, sales are high. Because most buyers finance their new home purchases, the monthly payment is as important to them as the total price if not more so. Purchasers will often defer their buying of a new home if they think that the interest rates are too high. Historically upturns in the interest rates and economic slowdowns have been difficult to predict. Local conditions (such as plant expansions, cutbacks, or closings) add to the complexity of the situation because they affect sales.

Establish a Budget

Tracking costs associated with development expenses is an important but a relatively easy task. As in any other form of cost control, the job begins with a budget for each cost category. This budget is derived from the estimate.

Set Up Cost Centers

In establishing the estimate, the first step is to determine your major cost categories. This following example uses the categories listed in Figure 7-2 (which were taken from Appendix F, Land Development Costs: Subsidiary Ledger, of *Accounting and Financial Management for Builders, Remodelers, and Developers*).[11] For your own purposes, you may want to change a few of these categories or add others.

Post Ledgers

Figure 7-3 shows a manually posted ledger for the costs associated with developing a 10-lot subdivision. Of course, this accounting could have been done more easily on a computer using a standard spreadsheet program or other suitable software (Figure 7-4).

For this particular project, the developer entered into an agreement to buy the raw land from John Landowner for a total price of $135,000 or $13,500 per lot. They agreed that the developer would pay $15,000 down and the additional payments would be made on a lot-release basis whereby the owner releases the lots one at a time for a payment of $12,000 each. By arranging the purchase in this way, the developer can borrow a smaller percentage of the total cost because the landowner is deferring the collection of his or her payment until the lots sell.

In this example, if the developer pays the $15,000 down payment out of his or her own funds and borrows the remaining money necessary for the development from a bank, the bank would be financing 54 percent of the cost. (A total cost of $295,100 minus $135,000 for the land leaves $135,000. Dividing that figure by the total cost is 54 percent). Of course, this practice benefits the developer because he or she does not have to borrow the $135,000 or pay interest on it.

Total cost	$295,100
Land cost	–135,000
	$160,100 ÷ $295,100 = 54%

In reviewing the costs associated with the job, note that the costs reflect the facts listed below:

- An independent land planner (A and B) was used to do preliminary layout and planning.
- Union Engineering did the design work to include market studies and permitting approvals.

Figure 7-2. Sample Land Development Costs — Subsidiary Ledger

0100 Preacquisition Costs
0101 Options
0102 Fees
0103 Professional services

0110 Acquisition Costs
0110 Purchase price, undeveloped land
0111 Sales commissions
0112 Legal fees
0113 Appraisals
0114 Closing costs
0115 Interest and financing fees

0120 Land Planning and Design
0121 Bonds
0122 Fees
0123 Permits

0130 Engineering
0131 Civil engineering
0132 Soil testing
0133 Traffic engineering

0140 Earthwork
0141 Fill dirt
0142 Clearing lot
0143 Rock removal
0144 Erosion control
0145 Dust control

0150 Utilities
0151 Sewer line
0152 Storm sewer
0153 Water line
0154 Gas lines
0155 Electric lines
0156 Telephone lines
0157 Cable TV lines

0160 Streets and Walks
0161 Curb and gutter
0162 Walkways
0163 Paving
0164 Street lights
0165 Street signs

1070 Signage
0171 Temporary
0172 Permanent

1080 Landscaping
0181 Sod or seed
0182 Shrubs
0183 Trees
0184 Mulch
0185 Other materials
0186 Other labor

0190 Amenities
0191 Swimming pool
0192 Tennis courts
0193 Tot lots
0194 Putting greens
0195 Exercise trail

National® Brand 45-621 Eye-Ease® Made in USA **Bally Acres**

	Initials	Date
Prepared By		
Approved By		

date	Payee	Ch #	1 Amount	2 Subtotal	3 Preacquisition Costs	4 Acquisition Costs	5 Land Planning & Design
			Budget	295,100	6,000	135,000	1,500
9 30	John Land Owner	640	15000 -	15000 -		15000 -	
10 21	A & B Land Planners	681	1550 -	16550 -	1550 -		
11 17	Union Engineering	737	1000 -	17550 -			
11 22	Soil Testing Inc.	753	485 -	18035 -			
12 13	Union Engineering	777	2750 -	20785 -			
12 15	Land Appraisal Corp	786	300 -	21085 -	300 -		
1 20	Dewey, Cheatham & Howe	920	200 -	21285 -	200 -		
1 20	Second National Bank	921	618 50	21903 50	618 50		
2 15	City of Gotham	973	1500 -	23403 50			1500 -
3 10	Earthmovers, Inc.	1011	8174 30	31577 80			
3 10	Second National Bank	1013	201 45	31779 25	201 45		
4 10	Earthmovers, Inc.	1071	7036 30	38815 55			
4 11	Sewer Builders, Inc.	1077	3500 -	42315 55			
4 11	Second National Bank	1079	251 69	42567 24	251 69		
5 10	Earthmovers, Inc.	1155	3125 07	45692 31			
5 10	Sewer Builders, Inc.	1156	48261 17	93953 48			
5 10	Second National Bank	1157	291 14	94244 62	291 14		
6 10	Sewer Builders, Inc.	1241	19436 90	113681 52			
6 10	Utility Constructors, Inc.	1242	21612 18	135293 70			
6 10	Second National Bank	1243	602 19	135895 89	602 19		
6 27	Street Builders Corp	1301	33705 27	169601 16			
6 29	Union Engineering	1311	1540 -	171141 16			
7 10	Riley's Landscape Service	1363	2000 -	173141 16			
7 10	W & F Construction	1364	4901 60	178042 76			
7 10	Second National Bank	1371	1289 51	179332 27	1289 51		
7 12	John Land Owner	1386	12000 -	191332 27		12000 -	
7 17	John Land Owner	1400	24000 -	215332 27		24000 -	
7 31	John Land Owner	1467	12000 -	227332 27		12000 -	
8 4	Earthmovers, Inc	1491	817 15	228149 42			
8 8	John Land Owner	1511	36000 -	264149 42		36000 -	
8 10	Second National Bank	1537	206 69	264356 11	206 69		
8 10	John Land Owner	1550	24000 -	288356 11		24000 -	
8 17	John Land Owner	1621	12000 -	300356 11		12000 -	
	ACTUAL EXPENDITURE			300356 11	5511 17	135000 -	1500 -
	VARIANCE			⟨5256 11⟩	488 83	0	0

Figure 7-3. Sample Manually Posted Job Cost Ledger

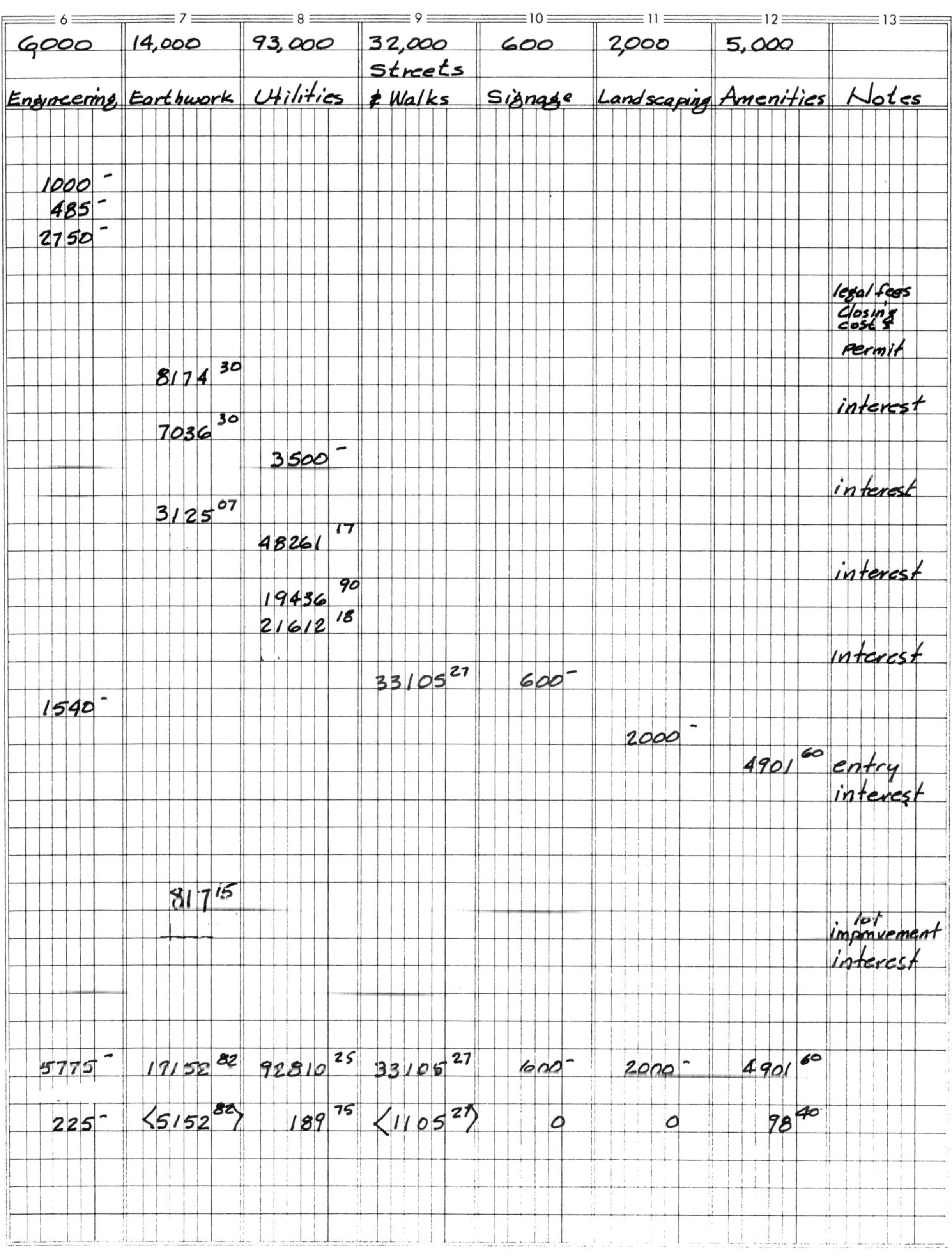

6	7	8	9	10	11	12	13
6000	14,000	93,000	32,000	600	2,000	5,000	
Engineering	Earthwork	Utilities	Streets & Walks	Signage	Landscaping	Amenities	Notes
1000 -							
485 -							
2750 -							
							legal fees
							Closing costs
							Permit
	8174.30						
							interest
	7036.30						
		3500 -					
							interest
	3125.07						
		48261.17					
							interest
		19436.90					
		21612.18					
							interest
			33105.27	600 -			
1540 -							
					2000 -		
						4901.60	entry
							interest
	817.15						
							lot improvement
							interest
5775 -	17152.82	92810.25	33105.27	600 -	2000 -	4901.60	
225 -	<5152.82>	189.75	<1105.27>	0	0	98.40	

Figure 7-3. Sample Manually Posted Job Cost Ledger (Continued)

Figure 7-4. Sample Computerized Job Cost Ledger

Itemized Category Report
9/1/93 Through 9/1/94

CONSTRUCT-All Accounts

Page 1

Date	Num	Description	Memo	Category	Amount
		INCOME/EXPENSE			
		EXPENSES			
		a Preacquisition Costs			
10/21/93	681	A&B Land Planners..		a Preacquisition Cost	- 1550.00
12/15/93	786	Land Appraisal Corp		a Preacquisition Cost	- 300.00
1/20/94	920	Dewey,Cheatham &...	Legal Fees	a Preacquisition Cost	- 200.00
1/20/94	921	Second National....	Closing Cost	a Preacquisition Cost	- 618.50
3/10/94	1013	Second National....	Interest	a Preacquisition Cost	- 201.45
4/11/94	1079	Second National....	Interest	a Preacquisition Cost	- 251.69
5/10/94	1157	Second National....	Interest	a Preacquisition Cost	- 291.14
6/10/94	1243	Second National....	Interest	a Preacquisition Cost	- 602.19
7/10/94	1371	Second National....	Interest	a Preacquisition Cost	- 1289.51
8/10/94	1537	Second National....	Interest	a Preacquisition Cost	- 206.69
		Total a Preacquisition Costs			- 5511.17
		b Acquisition Costs			
9/30/93	640	John Land Owner....		b Acquisition Cost	- 15000.00
7/12/94	1386	John Land Owner....		b Acquisition Cost	- 12000.00
7/17/94	1400	John Land Owner....		b Acquisition Cost	- 24000.00
7/31/94	1467	John Land Owner....		b Acquisition Cost	- 12000.00
8/8/94	1511	John Land Owner....	Lot Improvement	b Acquisition Cost	- 36000.00
8/10/94	1550	John Land Owner....		b Acquisition Cost	- 24000.00
8/17/94	1621	John Land Owner....		b Acquisition Cost	- 12000.00
		Total b Acquisition Costs			-135000.00
		c Land Planning, & Design			
2/15/94	973	City of Gotham.....	Permit	c LandPlanning & Design	- 1500.00
		Total c Land Planning & Design			- 1500.00
		d Engineering			
11/17/93	757	Union Engineering		d Engineering	- 1000.00
11/22/93	753	Soil Testing Inc..		d Engineering	- 485.00
12/13/94	777	Union Engineering		d Engineering	- 2750.00
6/29/94	1311	Union Engineering		d Engineering	- 1540.00
		Total d Engineering			- 5775.00
		e Earthwork			
3/10/94	1011	Earthmovers, Inc..		e Earthwork	- 8174.30
4/10/94	1071	Earthmovers, Inc..		e Earthwork	- 7036.30
5/10/94	1155	Earthmovers, Inc..		e Earthwork	- 3125.07
8/8/94	1511	Earthmovers, Inc..		e Earthwork	- 817.15
		Total		e Earthwork	- 19152.82
		f Utilities			
4/11/94	1077	Sewer Builders,Inc		f Utilities	- 3500.00
5/10/94	1156	Sewer Builders,Inc		f Utilities	- 48261.17
6/10/94	1241	Sewer Builders,Inc		f Utilities	- 19436.90
6/10/94	1242	Utility Constr....		f Utilities	- 21612.18
		Total f Utilities-			- 92810.25
		g Streets & Walks			
6/27/94	1301	Street Builders...		g Streets & Walks	- 33105.27
		Total g Streets & Walks			- 33105.27

Figure 7-4. Sample Computerized Job Cost Ledger (Continued)

Itemized Category Report
9/1/93 Through 9/1/94

CONSTRUCT-All Accounts Page 1

Date	Num	Description	Memo	Category	Amount
		h Signage			
6/27/94	1301	Street Builders...		h Signage	- 600.00
		Total h Signage			- 600.00
		i Landscaping			
7/10/94	1363	Riley's Landscaping		i Landscaping	- 2000.00
		Total i Landscaping			- 2000.00
		j Amenities			
7/10/94	1364	W&F Construction..	Entry	j Amenities	- 4901.60
		Total j Amenities			- 4901.60
		TOTAL EXPENSES			-300356.11
		TOTAL INCOME/EXPENSE			-300356.11

- A testing laboratory, Soil Testing Inc., did the preliminary soils analysis.
- Contracts for the construction of the development were let to Earthmovers, Inc.; Sewer Builders, Inc.; Utility Constructors, Inc.; and Street Builders, Inc..
- Riley's Landscape Service and W and F Construction did the landscaping and entry work.
- Interest was paid on a monthly basis.

Variance Between Budget Estimate and Cost

The ledger allows the developer to determine the variance between the estimate and the actual cost for each category. Although it will not necessarily tell why a variance occurred, it will show you where to start looking for the source of the problem. In this example the earthwork category ran over its budget by $5,152.82. Work such as earthwork, streets, and sewers is often done on a unit-price basis for which the engineer delineates the bid item and estimates the quantities for each. This method is commonly used when the exact quantities cannot be determined at the time of the bid.

A further examination of this example would show that a soft area not identified in the soils report had to be undercut and backfilled and that an additional $817.05 was spent to improve a lot as a part of the sales agreement. This same kind of examination would show that the second largest variance, an overrun of $1,105.27 in the streets and walks resulted from a design change to increase the pavement strength. Figure 7-5 provides a checklist for tracking and analyzing development costs.

Figure 7-5. Checklist for Tracking and Analyzing Development Costs

- Develop cost centers for your geographic area and level of work.
- Estimate the costs of a job and establish a budget.
- Explore cost-effective methods to do the work.
- Choose an engineer who is experienced in the exact kind of work you plan to do.
- Consistently compare your costs to the budget.

Notes

Introduction

1. Clough, Richard, and Glenn Sears. *Construction Contractor*, 6th ed. New York: Wiley and Sons, 1994, p. 336.
2. Jerry Householder, Chapter 4, "Preparing a Detailed Estimate," *Estimating for Home Builders* (Washington, D.C.: Home Builder Press, National Association of Home Builders, 1992), pp. 87–124.

2. Plan Your Approach to Cost Control

3. Chapter 4, "Preparing a Detailed Estimate," *Estimating for Home Builders* (Washington, D.C.: Home Builder Press, National Association of Home Builders, 1992), pp. 87–124.
4. Emma Shinn, *Accounting and Financial Management for Builders, Remodelers, and Developers* (Washington, D.C.: Home Builder Press, National Association of Home Builders, 1993), pp. 101–103.

Chapter 3. Track and Analyze Job Costs

5. *Software Directory for Builders and Remodelers*, 4th ed. (Washington, D.C.: Home Builder Press, National Association of Home Builders, 1995), 144 pp.
6. *Software Review: Approved Product Summaries for Builders*, 3rd. ed. (Washington, D.C.: Home Builder Press, National Association of Home Builders, 1994) 378 pp.
7. Construction Financial Management Association, Princeton Gateway Corporate Campus, 707 State Road, Suite 223, Princeton NJ 08540-1413, (609) 683-5000.

Chapter 4. Track and Analyze General Overhead Costs

8. Emma Shinn, *Accounting and Financial Management for Builders, Remodelers, and Developers* (Washington, D.C.: Home Builder Press, National Association of Home Builders, 1993), p. 101–103.

Chapter 5. Operate Your Purchasing System

9. Jerry Householder. *Scheduling for Builders* (Washington, D.C. Home Builder Press, National Association of Home Builders, 1990).
10. Construction Industry Institute, University of Texas at Austin, 3208 Red River, Suite 300, Austin, Texas 78705-2650.

Chapter 7. Track and Analyze Development Costs

11. Emma Shinn, *Accounting and Financial Management for Builders, Remodelers, and Developers* (Washington, D.C.: Home Builders Press; National Association of Home Builders, 1994), p. 115.

About the Author

Jerry Householder, builder, professor, and author, has many years of experience as a general contractor. He has built over $100 million worth of residential, commercial, and industrial projects throughout the southeastern United States.

Dr. Householder is Chairman of the Construction Department of Louisiana State University in Baton Rouge. Previous to that appointment, he was the Director of Graduate Studies in Construction Management in the College of Architecture and Urban Studies at Virginia Polytechnic Institute and State University. Dr. Householder teaches courses in planning and scheduling, construction management, and construction law. A professor for over 10 years, the author holds a doctorate in civil engineering from the Georgia Institute of Technology.

Dr. Householder wrote *Estimating for Home Builders* and coauthored *Basic Construction Management: The Superintendent's Job.* He also wrote the first edition of *Scheduling for Builders,* all published by Home Builder Press, National Association of Home Builders. He has written for the *ASCE Journal of Construction Engineering and Management, Arbitration Journal, Walker's Construction and Estimating,* and other periodicals. Dr. Householder also has presented seminars on building topics at the NAHB Annual Convention, and he frequently testifies as an expert witness in court cases involving disputes about construction jobs.

Increase Your Business Knowledge with These Bestsellers

Accounting and Financial Management, third edition—*Emma Shinn*— This revised and expanded edition shows you how to guide and evaluate your company's financial performance, design an accounting system, choose an accountant, and prepare, analyze, and use your financial reports. It includes new chapters for remodelers, developers, and multiproject companies; the complete, up-to-date NAHB Chart of Accounts; and a list of NAHB-approved software vendors.

Contracts and Liability for Builders and Remodelers—*NAHB Legal Department*—This expanded bestseller helps remodelers and builders avoid risks and protect against liability with well-written contracts. New chapters cover the contract between remodeler and owner, liability for builders and remodelers, design/build contracts used by remodelers and custom builders, and contracts with other team members.

Estimating for Home Builders— *Jerry Householder and John Mouton*—This book teaches you how to develop complete, accurate construction cost estimates and provides time-saving shortcuts. It gives detailed descriptions of all cost factors: subcontracts, materials, labor, tools, supplies and equipment, jobsite costs, overhead, and markup. Conversion tables, illustrations, and more.

Production Checklist for Builders and Remodelers—This step-by-step checklist helps you to get your projects done on time, within budget, and at high quality. It gives you day-to-day procedures to confirm start dates, monitor progress, predict completion dates, meet production schedules, and produce a high-quality home every time.

Residential Concrete, second edition—Everything you need to know about high-quality concreting—including guidelines for ordering ready-mixed concrete, working with admixtures, forming, jointing, and curing concrete. Shows you how to fix the most common concrete problems with detailed, easy-to-follow remedies.

Scheduling for Builders— *Jerry Householder*—Schedule your work more profitably with these easy-to-understand, step-by-step procedures. They will help you calculate construction timetables, forecast costs, and modify schedules when unexpected delays arise. Shows you how to streamline your scheduling process with bar charts, arrow diagrams, and precedence networks.

Understanding House Construction—*John Kilpatrick*—This book describes how homes are built from groundbreaking to final inspection. It introduces the methods, materials, and terms used in homebuilding and shows each step of the process with photos, diagrams, and explanations. Winner of the Distinguished Technical Communications Award presented by the Society for Technical Communications.

To order these books or to receive a current catalog of Home Builder Press products call (800) 223-2665 or write to—

Home Builder Bookstore®
National Association of Home Builders
1201 15th Street, NW
Washington, DC 20005-2800